THE SENTINEL STATE

THE SENTINEL STATE

Surveillance and the Survival of Dictatorship in China

Minxin Pei

HARVARD UNIVERSITY PRESS *Cambridge, Massachusetts* | *London, England* 2024

Printed in the United States of America

First printing

Library of Congress Cataloging-in-Publication Data

Names: Pei, Minxin, author.
Title: The sentinel state : Surveillance and the survival of dictatorship in China / Minxin Pei.
Description: Cambridge, Massachusetts ; London, England : Harvard University Press, 2023. | Includes bibliographical references and index.
Identifiers: LCCN 2023026654 | ISBN 9780674257832 (cloth)
Subjects: LCSH: Zhongguo gong chan dang—Discipline. | Intelligence service—China—History. | Dictatorship—China—History. | Secret service—China—History. | Information technology—Political aspects—China—History. | Technology and state—China—History. | Social control—China—History. | Spies—China—History. | China—Politics and government—1949–
Classification: LCC JQ1509.5.I6 P45 2023 | DDC 327.1251009—dc23/eng/20230713
LC record available at https://lccn.loc.gov/2023026654

To the memory of Pei Xingmei

Contents

Abbreviations

CCP	Chinese Communist Party, *zhongguo gongchandang*
CID	Central Investigation Department, *zhongdiao bu*
CPLC	Central Political and Legal Affairs Commission, *zhongyang zhengfawei*
DSP	Domestic Security Protection, *guobao*
FCE	Four Category Elements, *silei fenzi*
KI	Key Individuals program, *zhongdian renyuan*
KP	Key Populations program, *zhongdian renkou*
LSG	leading small group, *lingdao xiaozu*
MPS	Ministry of Public Security, *gong'an bu*
MSS	Ministry of State Security, *guojia anquan bu*
PLA	People's Liberation Army, *renmin jiefangjun*
PLC	political-legal committee, *zhengfawei*
PSB	Public Security Bureau, *gong'an ju*
PSC	Politburo Standing Committee, *zhengzhiju changwei*
PSD	Public Security Department, *gong'an ting*
SCS	Social Credit System, *shehui xinyong tixi*
SSB	State Security Bureau, *guo'an ju*

THE SENTINEL STATE

Introduction

The dystopian world portrayed in George Orwell's *1984* has long been regarded as the stuff of science fiction. But the Chinese government, with its aggressive adoption of the world's most advanced surveillance technologies, seems to be turning science fiction into reality. In a 2017 test of the potency of China's high-tech surveillance capabilities, BBC reporter John Sudworth challenged the police to find him while he walked the streets of Guiyang, a city of 5 million in Guizhou province. Armed with a photo of Sudworth, a local network of surveillance cameras, and facial-recognition technology, the police needed just seven minutes to locate him wandering the metropolis.[1]

Guiyang is like all Chinese cities, regardless of size: equipped with sophisticated technologies that sense and scan ordinary people, automatically capturing key identifying information that then populates and can be checked against police databases. According to a *New York Times* investigation published in late 2019, this omnipresent surveillance state can "help the police grab the identities of people as they walk down the street, find out who they are meeting with, and identify who does and doesn't belong to the Communist Party."[2]

Yet the fear and loss of privacy experienced by the average Chinese person are as nothing compared to the intrusions and humiliations suffered by members of the Uighur minority in the Xinjiang Uighur Autonomous Region. A *Wall Street Journal* reporter who traveled to the region in 2017 observed, "Security checkpoints with identification scanners guard the train station and roads in and out of town. Facial scanners track comings and goings at hotels, shopping malls and banks. Police use hand-held devices to search smartphones for encrypted chat apps, politically charged videos and other suspect content. To fill up with gas, drivers must first swipe their ID cards and stare into a camera."[3]

The power and reach of the Chinese surveillance state were on full display during the COVID-19 pandemic. To enforce its strict zero-COVID policy, the government used phone tracking, secret algorithms, and big data to ascertain ordinary people's health status and travel history and determine whether they would be allowed in public places or subject to quarantine. After spontaneous anti-lockdown protests broke out at the end of November 2022, police deployed the COVID-detection scheme to identify protesters who wore masks and eye goggles to thwart surveillance cameras and facial recognition—technologies that remained helpful in identifying protesters who went without similar precautions.[4]

The ruling Chinese Communist Party (CCP) does not seem to be satisfied with its surveillance power, however formidable it may appear to outsiders. It has announced an ambitious "social credit system" that could enable officials to assess an individual's political loyalty and to predict their intentions on the basis of data concerning their social and economic activities. The prospect of a "mother of all surveillance states" is so alarming that George Soros, the famed investor and open-society champion, warned of a scheme to "subordinate the fate of the individual to the interests of the one-party state in ways unprecedented in history."[5]

These accounts and cautions call our attention to the emergence in China of a technologically sophisticated surveillance system, but the public focus on the latest tools overlooks what might be more critical: the low-tech, labor-intensive approaches that lie at the foundation of the Chinese surveillance state. Over the course of decades, the Chinese party-state carefully assembled human resources and organizational architectures into which these technological factors have only recently been introduced. Indeed, the historical influences on the Chinese surveillance state date back a thousand years. It is my contention that the CCP's analog surveillance state, which has been under development since the Maoist era and became a focus of massive investment after the crushing of the Tiananmen pro-democracy movement in 1989, has been the key to the survival of the world's most powerful one-party dictatorship. Amid economic growth that, theoretically, should have facilitated liberalization or even democratization, the party-state has endured because it possesses the most capable surveillance infrastructure ever known.

The eye-catching technologies that the CCP has embraced with such enthusiasm are effective in no small part because they are deployed by the agents of security bureaucracies that are generously funded, well-organized, and carefully designed to deter and contain political threats to the party-state without themselves becoming threats to that state. This study aims to provide a comprehensive analysis of the evolution, organization, operation, and technological upgrading of China's surveillance state, to better understand both how dictatorships can endure and why other authoritarian regimes have faltered—sometimes at the hands of their own security apparatus—while the CCP thrived. By going back to the origins of modern Chinese surveillance architecture and observing how low-tech methods continue to function alongside new systems, we can better grasp how the CCP has successfully guarded its monopoly on power in spite of the revolutionary social and economic transformations of the post-Mao and post-Tiananmen eras.

The Study of State Surveillance in China

State coercion in China is a matter of considerable scholarly interest. Researchers have much to say about Chinese policing, in particular recent changes in the structure, resources, and tactics of the public-security apparatus as it seeks to enforce law and combat crime.[6] Scholars have also noted the priority given to security in the post-Tiananmen era. Since Tiananmen, officials in the public-security apparatus have gained elevated political status, and funding for domestic security has greatly increased.[7] Researchers describe the persecution of ethnic minorities and religious groups; violations of human rights; suppression of dissent; and "stability maintenance" (*weiwen*).[8]

Observers have also detailed an increasingly sophisticated menu of repressive tactics, including censorship of the internet generally and social media specifically, preemptive suppression of political activities during periods around sensitive anniversaries, the deployment of relatives and hired thugs to coerce citizens into ceding property rights or ceasing protest activities, the use of informants, infiltration of small villages by government agents, and the use of state welfare provision to keep tabs on ordinary Chinese.[9] Some of this research connects with China's adoption of advanced technology—the hot topic, popularized by journalists but also of great concern to scholars who have studied the matter in greater depth.[10]

With the possible exception of some flimsy journalism that gives excessive attention to technology, these studies have all been useful in advancing our understanding of the Chinese security state's tactics. But most leave aside the nature of the security state itself—its structure and organization. That there is a Chinese state security apparatus is taken for granted; the question is not what this apparatus is or how it works but what its effects are. Moreover, detailed research on specific instruments of repression may—perfectly understandably—have little to say about *why* the CCP is so heavily invested in domestic spying and social control.[11]

This book, informed by my own research and that of other scholars, tries to answer some of the more basic questions about China's surveillance state. Most importantly, I trace the evolution of the surveillance state and map its basic architecture. I offer in-depth investigations of the most critical components of China's surveillance state: the CCP's political-legal apparatus, which supervises and coordinates domestic security; the units in the coercive apparatus explicitly tasked with domestic spying; the network of informers; the main mass surveillance programs; and the most prevalent surveillance tactics.

Collectively, these structures and methods constitute the Chinese way of *preventive repression*. This concept will be central to the work ahead. Preventive repression is among the autocrat's most powerful weapons. The idea is simple: stymie the opposition before it can act. Preventive repression is not propaganda or ideological indoctrination; these aim to prevent the formation of an opposition in the first place. It is not just a matter of bribing potential enemies, either. And preventive repression is certainly not state violence in the form of arrests, beatings, imprisonment, and even the killing of dissidents. These are reactive techniques. Preventive repression, in contrast, reflects an understanding that the regime is best protected without eye-catching acts of violence. It subtly erects obstacles to opposition plans and in particular to collective action by those who would challenge and undermine the regime.

There are two reasons why China's contemporary surveillance state has so far escaped comprehensive examination. First, the party did not begin to expand, strengthen, and modernize its surveillance state until after the Tiananmen crisis in 1989, and a fully modernized surveillance state probably did not emerge until the end of the 2000s. As a result, there has not been much time to look back on the contemporary Chinese surveillance state and take stock of its development. Second, and more importantly, much of the information needed in order to understand the large-scale organization and operations of the surveillance state is not publicly available. One way around this obstacle would be to interview security agents in the field. But researchers have no easy access to such people and others with inside knowledge of the system.

Logistical challenges aside, an investigation that can pierce the veil of secrecy, shedding light on the surveillance state's structure, operation, and weaknesses, is of enormous value. Theoretically, such a study could contribute to the literature on the survival of authoritarian regimes in general, and on the relationship between regime type and durability in particular. As a case study, China's experience can illuminate how the institutions of a Leninist state equip its coercive apparatus with superior repressive and surveillance capabilities.

The Chinese case also adds nuance to thinking about the relationship between modernization and democratization. While wealthier societies tend to be more democratic, some dictatorships, most prominently China, are able to maintain their power despite sustained rapid economic growth. One likely explanation is that rapid growth increases the value of political monopoly, thus making these regimes even more determined to remain in power. At the same time, abundant revenue and access to advanced technologies enable these regimes to expand and upgrade their capacity to neutralize opposition.

Above all, this is an empirical study, providing careful and detailed description of the organization and operation of China's surveillance state. This descriptive project can help us reach our primary objective: to understand how the CCP has maintained its grip on power in spite of the massive socioeconomic changes China has experienced since the early 1990s.

Repression and the Survival of Dictatorships

All dictatorships, from patrimonial regimes, family dynasties, and military juntas to totalitarian, communist one-party states, depend on political repression to remain in power.[12] This is no easy task.

Dictators face two major challenges when relying on violence or threats of violence to deter and suppress opposition forces. The first is to fine-tune the amount of repression deployed. On the one hand, a dictatorship must repress its citizens: insufficient repression is unlikely to deter groups seeking to undermine its rule, thus imperiling its survival. On the other hand, excessive repression can incur severe reputational costs and international sanctions, while radicalizing opposition forces and precipitating violent backlash against ruling elites.[13] Excessive repression can make a dictator more insecure by alienating followers, whose fear for their own security may be an incentive to disloyalty.[14] Dictatorships deploying excessive repression tend to suppress economic growth, perhaps because repression often weakens property-rights protection and restricts economic freedom.[15] The political instability that excessive repression can trigger—backlash and even civil war—also depresses growth.[16] Over an extended period of time, lack of growth will cut into regime revenues, reducing the dictator's ability to buy support and fund the state's coercive apparatus, weakening the regime's hold on power. Solving the first dilemma of repression means discerning just how much repression is the right amount to shut down opposition without nourishing one's enemies.

The second challenge follows from the first. Repression—whether it is excessive or finely tuned—comes at the hand of state agents, usually quite a few of them. A dictator relying on repression typically must build and maintain a sizable coercive apparatus. But herein lies the "coercive dilemma," as it is known: the same apparatus that protects the ruler from popular uprisings may have the power to overthrow him.[17] Between 1950 and 2012, one-third of dictators removed from power were toppled by insider-led coups.[18] To counter this danger, dictators resort to coup-proofing tactics such as fostering rivalry among security forces. And by favoring one branch of the coercive apparatus over another—say, the secret police as opposed to the military—the dictator can prevent concentrations of power forming below him. But counterbalancing has costs. It can encourage disloyalty and degrade the effectiveness of targeted institutions. In addition, politicized security agencies may use excessive violence, provoking the first challenge discussed above.[19]

These problems are not insurmountable, but addressing them effectively is not simple. Usually a range of policies is needed, thus the most sophisticated and successful dictatorships—those that survive longest and rule in the most stable political environments—tend to have a diverse set of tools for maintaining power.[20] A dictatorship might promote economic growth and deliver a rising standard of living, using its performance on bread-and-butter issues to legitimize itself in the eyes of the public.[21] Using propaganda, education systems, and other means, dictatorships manipulate nationalist sentiment and public opinion generally in order to shore up popular support.[22] A regime may coopt selected social elites—business leaders, religious organizations, the intelligentsia, labor union bosses—by sharing its spoils; these elites then use their influence to broaden support for the regime.[23] Dictatorships skilled in practicing sham elections can add a veneer of democratic legitimacy to their claim to rule.[24] Such efforts come alongside repression, so that the latter is just one of the tools for retaining power, albeit an extremely important one.

Another further solution to coercive dilemma is to reduce the costs of repression by making it more efficient. A more efficient repressive apparatus does more for less: it maintains the rule of fear effectively without excessive violence and without augmenting the coercive apparatus to the point where it becomes a danger to the regime. In practice, the most efficient repression is preventive rather than reactive. Especially useful on this front is a highly organized and capable surveillance state that targets segments of the population most likely to lead or engage in antiregime activities, identifying and neutralizing opposition before it turns into collective action.[25]

Regimes that fail to maintain effective surveillance of opponents must resort to more costly measures: large-scale, long-duration imprisonment; torture, assassinations. These regimes may also be compelled to restrict basic freedoms and to stanch the flow of information so completely that markets become dysfunctional. Such overt and violent repression tends to inflame the opposition, which is precisely what the dictator does not want. In the worst case, a dictatorship with deficient capacity for preventive repression faces potentially dire consequences.[26] Massive street protests, such as those during the Arab Spring in 2010–2011 and the "color revolutions" in Georgia (2003), Ukraine (2004), and Kyrgyzstan (2005), can eject dictatorships.

Surveillance in Dictatorships

A dictatorship capable of organizing effective surveillance, such as that in East Germany (GDR) before the fall of the Berlin Wall and in China in the post-Tiananmen years, has the means to prevent and preempt organized opposition and collective action. At a minimum, surveillance allows the regime to gain timely and valuable intelligence about the activities and intentions of its opponents and to then take measures—such as issuing warnings and detaining key dissidents—to disrupt their plans. Another important function of surveillance lies in fostering restraint: simply knowing that secret police and their informants may be watching can lead one to curtail one's antiregime activities or take costly measures to evade authorities. Under certain circumstances, the understanding that one is or could be under surveillance may lead the opposition to abandon their efforts.

Besides prevention, effective surveillance produces spillover effects that further impede collective action on the part of the opposition. One of the key spillover effects is to induce in opponents a sense of fear and distrust. Consider that all dictatorships employ spies and informants recruited from close associates—mostly colleagues and neighbors—of known or suspected opponents. These agents keep the secret police informed, but their value exceeds these direct contributions. The possibility of infiltration also breeds suspicion among opponents, who worry that their associates are working for the dictator. The resulting distrust, even paranoia, renders collective action more difficult because dissidents do not know whom to trust with their plans. In a society with a large number of informants, recruiting dissidents is difficult because individuals approached for participation in politically sensitive activities are naturally fearful of entrapment and will be reluctant to commit. Thus informants not only provide intelligence, they also, by virtue of their mere presence or even potential presence, prevent the opposition from forming, expanding, and organizing.

Dictatorships of all stripes typically entrust the preventive repression of domestic opposition to a specialized bureaucracy—usually the secret police, who conduct surveillance by monitoring communications and mobilizing informants.[27] It is easy to understand the rationale behind tasking the secret police with surveillance. As surveillance is typically covert, the bureaucracy implementing this task must be shrouded in secrecy. Additionally, because most surveillance activities require specialized skills—such as running informants, infiltrating opposition groups, and operating sophisticated eavesdropping equipment—the regular police may not be up to the task. The secret police, then, will tend to vet their recruits more rigorously than regular police do and will enjoy a more elite status and greater trust from the ruler.

While secret police can be highly effective, granting them exclusive responsibility for surveillance also carries significant risks.

Their elite status and power can easily result in corruption. A powerful domestic spying agency can become a state within a state, with its own bureaucratic interests and an agenda that may diverge from that of the ruler. When ambitious spy chiefs control the secret police for an extended period of time, they often turn the secret police into their own personal fiefdoms and threaten the power of their political masters.[28]

Another serious drawback of employing of secret police is the cost. Since nearly all dictatorships rule low-to-middle-income countries, they can hardly afford to employ a large secret police force. For example, the State Intelligence and Security Organization (SAVAK) under the shah of Iran had only 5,300 employees and 55,000 informants in the 1970s, when Iran had a population of 33 million. That is one SAVAK employee for every 6,200 people and one informant for every 550 people. By comparison, the elite police agency of the United States, the Federal Bureau of Investigation, has only about 35,000 employees, one for every 950,000 or so Americans. The Dirección de inteligencia nacional, the secret police under Chile's military dictatorship, reportedly had only 2,000 operatives for a population of well over 10 million.[29] In the former Soviet Union, the ratio of KGB officers to the population was estimated to be one per every 595 people. The equivalent ratio was one per every 1,553 in Romania, one per every 867 in Czechoslovakia, and one per every 1,574 in Poland. East Germany was the exception that proves the rule. Its Ministry of State Security (MfS, or Stasi) had 91,015 full-time agents in 1989, one agent per every 165 people.[30] In addition, the Stasi had 189,000 informants at the time of the fall of the Berlin Wall, or more than one in every 100 people. In total there were about 600,000 Stasi informants over the course of East Germany's existence as a state.[31] In most low- or middle-income countries, the resources required for keeping a secret police apparatus as large as the Stasi would be unthinkable. In the case of China, employing one secret police agent for every 165 people would require 8.5 million secret police officers—more than four times the size of its ordinary police force, per a 2010 statement from the minister of public security.[32]

The alternative is to recruit a large number of informants, but this approach has its own limitations. Although informants can expand the intelligence capabilities of the secret police at less cost than hiring and training secret agents, financial and operational constraints do cap the number of informants that can be productively employed. Unless ideologically or politically motivated, most informants must be paid; for instance, informants in Iraq under Saddam Hussein received a monthly stipend.[33] East Germany gave informants permission to travel abroad as well as preferential access to housing, cars, telephones—as long as they were available.[34] In communist regimes, which dominate the economy and control key social institutions such as universities and cultural organizations, secret police can leverage their influence over employment and licensing for recruitment purposes, helping them maintain a large stable of informants. But most authoritarian regimes lack such capacity.

It is also the case that more informants are not necessarily better, as an officer of the secret police may not be able to effectively manage a large number of them. In addition to vetting and training, the secret police must maintain regular contacts with informants and evaluate the intelligence they provide.[35] Such operations are time-consuming and labor-intensive. Meeting with an informer typically requires the presence of two secret police officers, to ensure accurate notetaking and hedge against the possibility that secret police are themselves untrustworthy. If we assume that only half of the SAVAK's 5,300 full-time agents in the 1970s supervised their 55,000 informants, then each supervising officer would have twenty informants reporting to them. In East Germany, the ratio of Stasi agents to informants was roughly one to two. If only a quarter of Stasi agents had this responsibility, their task would have been far more manageable than the SAVAK's.

As most dictatorships are incapable of maintaining a large, well-resourced secret police agency like the Stasi, the scope and intensity of surveillance by secret police is necessarily limited. Consequently, preventive repression—and, indeed, surveillance itself—cannot depend solely on secret police. A broader surveillance state turns out to be critical.

Organizing Surveillance

The term surveillance state is widely used but rarely defined. In this study, I define the surveillance state as a system of bureaucratic agencies, human and technological networks, and state initiatives undertaken for the purpose of gaining intelligence on the public's activities, private communications, and public speech, especially that of individuals and organizations deemed threats or potential threats to the ruling elites. On this view, the surveillance state comprises individual components, each of which can be analyzed in relation to one another and in terms of their operational tactics and various other factors influencing their effectiveness. Yet while the components of a surveillance are functionally distinct, they all serve a common objective: preventive repression through the collection and mobilization of intelligence concerning domestic political opponents.

The effectiveness of a surveillance state rests not only on the capabilities of each component but also on the coordination and integration of these components. Thus the secret police, ordinarily the most crucial component of any surveillance state, is likely to be most effective when it enjoys cooperation from other parts of the state. Simply put, when it comes to a surveillance state, the whole is more than the sum of the parts. In consequence, any inquiry into the defense of dictatorships should pay close attention to the ways in which surveillance activities are—or fail to be—harmonized across bureaucratic organizations, including but not limited to secret police. Because "siloed" bureaucratic departments make coordination of surveillance difficult, effective coordination typically requires that the departments be arrayed under a separate umbrella organization that embodies undisputed authority and can reward cooperation and impose penalties in order to align activities across the organs of the surveillance state.

As a result, the most effective surveillance states should be those that face the fewest constraints in terms of political empowerment, organizational capacity, and material resources, including funding and technological capabilities. Political empowerment refers to the priority assigned by the ruler to preventive repression and the authority and discretion granted to the organizations tasked with surveillance. In this respect, the contrast between democracies and dictatorships is truly stark. Law enforcement agencies in democracies possess potent surveillance capabilities, but, because democracies protect citizens' rights to privacy, these agencies lack the authority and discretion enjoyed by their counterparts in dictatorships. There are differences even among autocracies: state surveillance in soft authoritarian regimes—such as Mexico under the Institutional Revolutionary Party, Malaysia under the United Malays National

Organization, and Singapore under the People's Action Party—has been relatively unintrusive.[36] In addition, the priority an autocracy gives to preventive repression may differ from one period to another. As we will see, the surveillance state in China operated under significant political constraints in the 1980s thanks to liberal reformers. Two in particular, Hu Yaobang and Zhao Ziyang, ran the CCP's day-to-day affairs and used their positions to keep the security apparatus on a tight leash.

Alongside political empowerment, organizational capacity is critical to effective surveillance. Regimes capable of building and maintaining a complex, competent, and politically loyal surveillance apparatus will succeed in preventive repression where others fail. In particular, dictatorships enjoying a high degree of *regime penetration*—the ability to reach and control key economic and social institutions as well as grassroots society—are able to recruit more informants and activists who can assist the secret police. Societally embedded surveillance provides the coercive apparatus with sources of low-cost intelligence and the means to closely monitor and intimidate known and suspected threats. Regime capacity also fosters mechanisms for coordinating the various components of the repressive apparatus. Such mechanisms may include the aforementioned executive body overseeing all aspects of surveillance and domestic security; specialized agencies or offices that bring together staff from across relevant government agencies to handle a single top-priority task; and routine cross-agency meetings in which the regime communicates its security agendas and assorted components of the coercive apparatus exchange knowledge.

Finally, a dictatorship's access to fiscal resources and advanced technologies obviously can strengthen its surveillance capabilities.

However, material resources alone are insufficient. Without adequate organizational sophistication and political oversight, money and technology are likely to be wasted, misused, or lost to corruption. Only the combination of resources and organizational capabilities can enable a dictatorship to build a fully equipped surveillance state. In this regard, China in the post-Tiananmen period has been truly exemplary.

The Leninist Surveillance State

Even a casual look at the determinants of effective surveillance will show that they are beyond the grasp of most dictatorships due to their weak organizational sophistication and lack of resources. This should not be surprising, since a surveillance state is essentially a state in miniature. Some dictatorships may be exceptionally ruthless, but these are unlikely to possess a capable surveillance state if their state institutions are poorly organized and possess weak capacity.

The connection between state capacity and the effectiveness of surveillance states explains why personal dictatorships, however brutal, are rarely capable surveillance states. Suharto's Indonesia, Iran under the shah, Haiti under the Duvaliers, and Ferdinand Marcos Sr.'s Philippines are but a few well-known examples. As personal dictatorships typically centralize power in the hands of a ruler who distrusts state institutions that can threaten his power, these regimes control relatively weak states and rely on politicized coercive apparatuses characterized by patronage, rivalry, and poor technocratic capabilities.[37] Additionally, as personal dictators have few ties to society and govern mainly through patrimonial arrangements—that is, by doling out favors to individuals and groups in exchange for loyalty—they are incapable of embedding their repressive apparatus in society or mobilizing societal resources to augment surveillance capabilities. Some personal dictatorships, like the shah's or that of Hosni Mubarak in Egypt, possess fearsome secret police.[38] However, even the SAVAK and Egypt's much-feared Mukhabarat could not match the surveillance capabilities of their counterparts in communist states such as the Soviet Union, East Germany, and today's China, because the Iranian and Egyptian regimes lacked the capacity to penetrate society and embed large surveillance networks into the workings of daily life.[39] The Iranian and Egyptian regimes were thus fairly effective in terms of reactive repression, but less so preventive repression.

Military regimes may be only marginally more effective than personal dictatorships in building and maintaining surveillance states. Like personal dictatorships, military regimes are hampered by isolation: personal dictators are constrained by their narrow social base of support and lack of state institutions capable of carrying out their will, while military dictators are constrained by the institutional insularity of the armed forces, which limits the regime's capacity to coordinate policy across state bureaucracies, build durable ties with major social groups, and mobilize societal resources for coercion and surveillance. Military regimes also tend to underinvest in surveillance out of a (largely justified) belief that they can count on their troops to crush any threats to their power.[40]

Perhaps the most capable dictatorship, from the standpoint of installing a surveillance state, is the Leninist state. Leninism here does not refer to the ideological characteristics of Vladimir Lenin's thought, which tended mostly toward orthodox Marxism. Rather, I have in mind the organizational structure of the one-party state that he and his cadre pioneered in the early days of the Soviet

Union. Leninist regimes are ruled by a highly institutionalized party with a strict organizational hierarchy, established appointment and promotion procedures, and extensive societal links through the party's control of critical sectors, such as the economy, education, science, and culture.[41] These regime features are tailor made for surveillance. What is more, despite their elitist nature, Leninist parties maintain an on-the-ground presence in the form of party committees and cells in all key state, social, and economic institutions.[42] Even after the totalitarian phases of the Soviet Union and the People's Republic of China—which ended with the deaths of Joseph Stalin in 1953 and Mao Zedong in 1976, respectively—their Leninist communist parties remained dominant, proving their durability even in the absence of charismatic leadership and violent repression.[43]

The penetration of the state, society, and the economy by a Leninist party is the key to establishing favorable conditions for a highly sophisticated surveillance state. Politically, the supremacy of the party ensures the loyalty of the repressive apparatus and implementation of the party's security policy through personnel appointments, promotions, and material incentives. Organizationally, the top-down Leninist system is suited to communicating the leadership's domestic security priorities, translating them into policies, and overseeing their implementation. Operationally, the omnipresence of the party in the institutions of the national state, local states, major businesses, and social institutions such as universities facilitates the coordination of surveillance and mobilization of the administrative and material resources available to build and maintain the surveillance state. Leninist regimes' capacity for coordination may give them an edge in the adoption of surveillance technology as well. Even though these regimes are not necessarily or especially suited to technological innovation, they can use their organizational reach and regulatory power to order local governments, state-affiliated entities, and private businesses to install surveillance technologies quickly and at scale.

The unrivaled social and economic penetration of Leninist regimes also promotes recruitment of informants. Secret police can coerce potential recruits by threatening them with loss of jobs, licenses, and other benefits if they refuse to spy on their fellow citizens. In addition, Leninist regimes also can call on party members to act as informants or perform security duties at no direct cost. In East Germany, for example, one in twenty Communist Party members was a Stasi spy. About half of the members of China's local security committees in the 1960s were members of the CCP and its affiliated youth league. Although the government has not made public more recent data, references to the mobilization of party and Youth League members for volunteer security activities can be readily found on websites of local governments and universities.[44]

Furthermore, a Leninist party-state can defray costs of the surveillance state in a way other dictatorships cannot. Most dictatorships can pay for surveillance using only the state funds available for policing. In China, matters are different because, per the Leninist model, party officials are embedded in the leadership of essentially all institutions of any scale or importance. The officials have direct control over universities, businesses, and other nongovernmental institutions. They also direct the expenditures of local governments. Thus, any pot of money, associated with any institution, might be funding surveillance. As Chapter 4 shows, local governments bear the costs of recruiting and maintaining a network of informants much larger than the one operated by the police. And as detailed in Chapter 5, local government informants and officials surveil a larger number of targeted individuals than police do.

Surveillance with Chinese Characteristics

Postrevolutionary China combines the unrivaled capacities of a Leninist party-state with elements of a system of social control dating back a thousand years. The CCP surveillance apparatus draws on one of the institutional legacies of China's imperial period, the *baojia* system. Introduced in the eleventh century by Wang Anshi, the reformist prime minister of the Northern Song dynasty, the baojia combined elements of urban planning, census taking, tax collection, and law enforcement to enforce social order at the local level. Households were parceled ten at a time into a *bao,* with ten baos grouped into a *jia.* Each household reported problems in the community—crimes, conflicts, or suspicious activities—to their bao leader, who might then report to the jia leader. In turn, a jia leader went up the chain of command to state authorities. Families had strong incentive to monitor and report, as they were held responsible for crimes or other misconduct committed by their neighbors.[45] Households were also registered for purposes of levying taxes and impressing labor, and the state collected information on family composition.

The baojia was not fully implemented; some conservatives opposed Wang's ambitious reform, and much of the population resented the imposition of hierarchical, militarized organization on daily life.[46] During the Ming Dynasty, the baojia was practiced in some areas but was not implemented nationwide.[47] But the Qing Dynasty resuscitated the baojia and restored its surveillance function.[48] The baojia even survived the collapse of the Qing Dynasty in 1911. Japanese colonizers applied it in Manchuria and Taiwan, as did the Chinese Nationalist government in 1939 an effort at state-building.[49]

In contemporary China, a modernized baojia operates in the forms of the *hukou* (household registration) and *wanggehua guanli* (grid management). The hukou collects information on individuals—their gender, date of birth, national identification number, ethnicity, religion, place of residence, level of education, employment status, blood type, marital status, and so on. Since the founding of the People's Republic in 1949, an individual's status as a rural or urban resident has been used in determining whether they will receive state benefits such as access to publicly funded education and other social services. Grid management, meanwhile, is an evolution on the neighborhood divisions of the baojia. Population centers are sliced into sectors for purposes of public safety, improving state services, controlling traffic, and handling sanitation.

The combination of the hukou and grid management—supplemented by modern technology—is far more powerful than the baojia. During the imperial period, the state lacked the coercive and organizational resources to turn the baojia into a really capable surveillance tool. But the Leninist party-state possesses such resources in abundance and has penetrated society deeply and extensively, not least through party cells and members distributed across cities large and small. At the end of 2021, the CCP had 96.7 million members and 4.93 million local branches.[50] Relying on its members and activists, the party has formed quasi-state organizations such as neighborhood and village committees that enforce social control with the aid of grid management. Community police stations, absent in imperial China, are now ubiquitous and function as a vital cog in the machinery of state surveillance. Most critically, the imperial state could not use the baojia to control access to employment, food, housing, and social services because the state did not dominate the economy. By comparison, China under Mao and since has relied on the hukou to do precisely this. Through these measures, the contemporary Chinese state not only oversees daily life but also gains granular information about the population.

To ensure that this structure continually provides authorities the information they need in order to maintain social control, the Chinese Leninist party-state has developed, mainly through a process of trial and error and adaptation, an innovative system that I call *distributed surveillance*. The essence of distributed surveillance is to spread the responsibilities and costs of surveillance across various security bureaucracies, other state actors, and nonstate actors, with coordination performed by a specialized party bureaucracy—the CCP's political-legal committees.

Responsibility for surveillance is divided among three police bureaucracies: the Domestic Security Protection unit within the Ministry of Public Security, frontline police stations, and provincial and municipal state security bureaus. Importantly, these agencies are not organized hierarchically. Rather, each has its separate areas of jurisdiction, not unlike local, state, and federal law enforcement agencies in the United States. In this way, the party-state avoids concentrating power in a single security bureaucracy. Such distribution of responsibility differs from the classic coup-proofing strategy of counterbalancing in that a relatively clear division of labor among these security bureaucracies prevents rivalries that could impair their operational effectiveness. Additionally, the party assigns secondary surveillance tasks to other state actors—state-affiliated institutions such as business enterprises, universities, local neighborhood and village committees—and collaborators, including party loyalists and informants.

Distributed surveillance not only prevents concentrations of power within the security apparatus, it also helps contain costs by shifting a significant burden to state-affiliated nonsecurity entities and low-paid or unpaid informants. The marginal cost of secondary surveillance is low from the perspective of businesses and universities, because their personnel would be in place even in the absence of their surveillance responsibilities; these are added tasks, but they require no additional manpower and consume relatively little staff time. For their part, informants are often paid only when they produce valuable intelligence and may be remunerated with nonmonetary rewards.

Overseeing distributed surveillance are the party's political-legal committees, which enforce political loyalty and coordinate the surveillance activities of state and nonstate actors. Such an umbrella structure is possible thanks to the party-state. The Leninist ideal is a central authority embedded at every scale of society, with agents and informants operating in key economic entities, such as large state-owned enterprises; major social institutions, including universities; and down to the neighborhood level. The party-state thus is intimately present in the various contexts of daily life, shaping access to opportunities, observing people's behavior, and collecting information.

The Surveillance State and the China Puzzle

The Chinese experience in the post-1989 period poses a frontal challenge to the long-standing and influential theory that economic modernization is associated with democratization.[51] The idea is that economic growth brings with it democratic values, deconcentration of resources, and a growing middle class and civil society capable of organizing resistance to state oppression. A population with expanded access to information gains capacity for collective action and demands a greater voice in governance. How has China's one-party dictatorship managed to thrive amid sustained economic development?

Many explanations for the Chinese party-state's survival have been advanced. Some attribute its durability to success in coopting social elites, in particular private entrepreneurs. Others credit the party's adaptive capabilities. The CCP is seen as a successful institutional reformer, managing internal conflict and meeting public demands by imposing term limits, establishing meritocracy in appointment and promotion of party and government officials, and improving state responsiveness to the population's everyday needs. Overall, the public has been satisfied with the economic performance of the CCP, which is surely important to the regime's survival.[52]

A plausible explanation of the China puzzle is that the evolving surveillance state has enhanced the CCP's capacity for preventive repression to such an extent that the party-state is able to neutralize threats created by economic reform and modernization.[53] The improvement in the CCP's preventive repression has been remarkable, enabling it to avoid the most brutal methods of deterring political opposition and collective action (except in Tibet and Xinjiang). While human rights abuses are routine and widespread, the number of political prisoners—a measure of brutality and a barometer of reactive, as opposed to preventive, repression—has been relatively small in the post-Tiananmen era.[54]

Autocrats ruling fast-growing economies have both the incentives and the means to strengthen their surveillance capabilities.[55] Scholars of democratic transition have made a compelling argument that transition to democracy in most authoritarian regimes is fundamentally a decision made by the ruling elites. Typically, a crisis of legitimacy compels autocrats to liberalize.[56] By contrast, a dictatorship presiding over a fast-modernizing society can claim performance legitimacy and avoid such a crisis. More importantly, rising prosperity increases the value of political monopoly as autocrats can convert power into wealth (or reap more lucrative "rents"). Consequently, these autocrats become even more determined to defend their power instead of initiating liberalizing reforms.[57]

Unlike their counterparts in economically failing dictatorships, autocrats in control of prosperous economies have ample resources to expand and upgrade their surveillance capabilities by increasing the size of the secret police, recruiting a large number of informants, targeting more groups, and acquiring advanced technologies. A dictatorship able to utilize the rising wealth of modernization to strengthen its surveillance capacity is obviously less likely to fail.[58]

Objectives and Arguments

This book has two objectives. Empirically, it seeks to map the architecture of China's surveillance state, uncover its programs and operational tactics, and argue that what makes this surveillance state uniquely formidable is the combination of a Leninist regime's organizational capacity and the more recent acquisition of advanced technologies. Theoretically, this book contributes to the literature on state surveillance in dictatorships by exploring the role of distributed surveillance in resolving the coercive dilemma and the practical challenges of state surveillance. I also provide a more narrowly focused explanation for why economic modernization may perversely, at least in the short-to-medium term, enable a dictatorship to better guard itself with more effective surveillance capabilities. Simply put, successful economic modernization, as in post-1989 China, not only makes autocrats more determined to defend their power but also generates the essential resources for them to strengthen their capacity for preventively repressing those pro-democracy forces that have been produced by modernization. The implication of this argument is that democratic transitions are less likely to occur in dictatorships that succeed in modernization than in those that fail in such endeavors. In the Chinese context, it is economic failure, not success, that will more likely be the precursor of a future transition to democracy.

As the primary contribution of this study is empirical, the book traces the evolution of China's surveillance state, maps its organizational architecture, and describes its operational tactics. Chapter 1 reconstructs the historical development of the surveillance state. Chapter 2 focuses on the institutions and mechanisms that coordinate surveillance. Chapter 3 describes the organizational structure of China's multilayered surveillance apparatus. Chapter 4 probes the country's vast network of spies and informants. Chapter 5 examines two mass surveillance programs targeting individuals. Chapter 6 analyzes surveillance tactics, and Chapter 7 describes the technological upgrading of surveillance since the late 1990s.

Before turning to these issues in earnest, one clarification may be necessary: surveillance may seem conceptually distinct from repression, but this distinction is often blurred in reality. Surveillance conducted for the purposes of preventing or suppressing dissent may not involve violence or explicit coercion, especially when surveillance operations are concealed from their targets. But nonviolent surveillance should be considered repressive, as its main purpose is to prevent or stifle peaceful challenges to the authority of a dictatorship. Operationally, parts of the coercive apparatus performing surveillance are also the same police agencies that carry out repressive activities such as intimidation, harassment, and arrests. Consequently, the surveillance state is embedded in the coercive apparatus.

I previously noted that direct documentation of the Chinese surveillance state is difficult to come by. This is true, yet many secrets of the Chinese surveillance state are hiding in plain sight. This study is made feasible by a trove of official sources—mainly local yearbooks and gazettes—that disclose brief, but crucial, details about the organization and operation of China's surveillance state. In some cases, leaked classified materials also supply valuable evidence. On occasion, information posted on official websites inadvertently reveals useful clues as well. Researchers can fruitfully exploit the leakages in the Chinese system. To be sure, this book leaves crucial questions unanswered. For example, available sources tell us little about the operations of the Chinese surveillance state in ethnic minority areas such as Tibet and Xinjiang. I am able to shed light on surveillance of Tibetan Buddhist monasteries but otherwise do not focus on these two sensitive regions. Indeed, information is not the only obstacle. Given the unique ethnic, economic, social, and geographic conditions of these two regions, a standalone research project on the Chinese surveillance state in Tibet and Xinjiang would be more appropriate, if and when information becomes available. Still, my hope is that this investigation will make significant progress in understanding China's surveillance apparatus and its role in preserving and further empowering a regime that, theoretically, should by now have been transformed into a more open one.

CHAPTER 1

The Evolution of the Chinese Surveillance State

Four distinct phases mark the evolution of China's surveillance state. During the first phase, in the decade after the revolution (1950–1959), the basic framework of surveillance emerged in the context of a totalitarian institution-building project. In the second phase, from the Great Leap Forward in 1959 to the end of the Cultural Revolution in 1976, economic and political shocks devastated the surveillance state. During the third phase, beginning with the opening of reform in 1979 and ending with the Tiananmen crackdown in 1989, the surveillance state was gradually repaired and professionalized. However, the restored surveillance state lacked a clear political mission: the domestic-security apparatus devoted most of its resources to Deng Xiaoping's "strike-hard" (*yanda*) anticrime campaigns, and the generally tolerant political atmosphere and presence of reformers in party leadership limited the use of preventive repression. Resource constraints also impeded modernization of the surveillance state. Almost everything changed in the post-Tiananmen era, the fourth phase. The party's near-death experience in 1989 gave the leadership an existential incentive to expand

and upgrade surveillance through investment, institutionalization, and technology.

We can make two broad claims about China's surveillance state in the post-Mao era. First, transformative economic modernization during this period did not touch the basic institutional and organizational frameworks of a totalitarian surveillance state. On the contrary, resources generated by rapid economic growth allowed the party-state to strengthen its coercive capacity. Second, a well-resourced and technologically advanced surveillance state did not emerge until the post-Tiananmen era.

The Surveillance State in the Maoist Era

During their first decade of rule, Mao and his fellow revolutionaries focused heavily on consolidating power by eliminating potential domestic enemies, labeled as counterrevolutionaries. These were loosely defined as spies, key figures in reactionary parties, leaders of reactionary sects and secret societies, landlords, and officials of the Kuomintang and Japanese puppet regimes who continued to maintain a "reactionary stance."[1] The 1950s saw nationwide campaigns of terror, intended to identify political opponents and execute or imprison the most dangerous among them. But if the terror campaigns were a momentary response to fears of opposition, the party-state also developed organizational structures for enforcing social control over the long term: an expanding police force; a network of spies and informants that penetrated to the grassroots level; local security committees that could, thanks to the ideological zeal of their members, be trusted to operate with discretion; the beginnings of a centralized security bureaucracy and policymaking apparatus; and durable programs of mass surveillance.

All of these efforts faced a key structural challenge: China is a vast and enormously populous state, making coordination of security agents difficult. Then too, the early postrevolutionary state was neither wealthy nor technologically sophisticated. A major part of the solution was mass mobilization: relying on ordinary citizens to monitor and report on each other. Still, erecting the surveillance state took time, and the process was uneven, with quicker gains made in urban than in rural areas. Furthermore, the very same ideological zeal that underlay mass mobilization could be at odds with security goals. Top-down projects of social and political transformation—like the Great Leap Forward, Anti-Rightist Campaign, and Cultural Revolution—all stoked public support for the regime. But the associated purges—and the famine caused by the Leap and political havoc of the Cultural Revolution—undercut the nascent security state. It did not begin to recover until the post-Mao era.

Terror Campaigns

Like other totalitarian regimes, the new Chinese party-state resorted to mass arrests, imprisonment, and executions to establish a rule of fear and to identify political threats.[2] Between October 1950 and the end of 1955, the CCP orchestrated terror campaigns in which 4.6 million people were arrested, out of a population of just under 600 million.[3] Among these, 770,000 were executed and 1.8 million imprisoned. In addition, 600,000 counterrevolutionaries were placed under *guanzhi* ("enforced control"). While they were not incarcerated, they lost much of their freedom: subjects of guanzhi require authorization to do even day-to-do tasks and are required to report on their activities.[4] In addition, the regime required that all members of "reactionary parties and groups" register with the local authorities. Provincial public-security gazettes reveal that hundreds of thousands of "key elements" in these reactionary parties and groups turned themselves in.[5]

After the registration campaign was completed in 1952, the party banned "reactionary sects and societies," which had vast memberships and could become organized threats. Jiangxi's Public Security Department (PSD) claimed that, in 1949, there were thirty-two major sects and societies in the province, with close to 14,000 leaders and 210,000 members.[6] In Shanghai, the municipal Public Security Bureau (PSB) designated 52 of 203 known sects and societies "reactionary."[7] During the crackdown, leaders of these sects and societies were arrested and executed. In Jiangxi, police arrested 907 sect leaders and forced more than 123,000 members to renounce their affiliations. Sixty-five such leaders were executed in Shanghai, and more than 320,000 people were forced to renounce their sect membership. Police in Zhejiang Province took credit for banning 264 sects and secret societies, arresting 2,136 leaders, and making 649,200 people renounce their memberships.[8]

Yet even mass terror failed to ease the regime's insecurities. In July 1955, fearing that a large number of dangerous elements had evaded the dragnet, the party ordered a new campaign to "purge concealed counterrevolutionaries." The operating assumption was that concealed counterrevolutionaries and other bad elements constituted 5 percent of all people involved in government agencies, the military, schools, and enterprises.[9] When the campaign ended in October 1957, 18.45 million people had been vetted and 100,232 were uncovered as hidden counterrevolutionaries and bad elements.[10] (The police formally recognized the category of "bad elements," albeit that this group was ill-defined, as I discuss below.)

By March 1959, when the successive terror campaigns were finally over, the party had achieved most of its key objectives.[11] In addition to destroying the bulk of the elites from the old regime and potential opposition leaders, it had identified millions of other apparent political threats and had firmly established a rule of fear. In September 1957, Luo Ruiqing, chief of the Ministry of Public Security (MPS), estimated that only 2 percent of the country's population could be considered part of the "counterrevolutionary social base." Still, that amounted to some 12 million people, who subsequently became targets of state surveillance.[12]

The Police Force

Except during a brief period between 1986 and 1991, the Chinese government has never published the size of its civilian police force, officially called the People's Police. Nevertheless, we can estimate the growth of China's police force based on disclosures in local public-security gazettes and in the internal MPS publication, *Major Events in Public Security since the Founding of the PRC.*[13] (These estimates exclude the People's Armed Police, a domestic paramilitary unit that, among other tasks, guards foreign diplomatic missions, combats terrorism, and responds to mass riots.) The MPS reports that, in late 1958, China had 400,800 police.[14] By June 1984 the government-authorized force totaled 658,000.[15] This reflects the number of officers approved for employment, but assuming this was roughly the actual number of police—a safe bet—the size of the formal police force increased by about 2.5 percent per year between 1958 and 1984, roughly in line with population growth, which was 2.3 percent per year over the same period.[16]

The growth was not, however, linear—thanks to the twin shocks of the Great Leap Forward (1958–1959) and the Cultural Revolution

(1966–1976). The former, an ill-conceived economic and social transformation launched by Mao, led to economic collapse and the worst famine in modern history, with tens of millions dead. The latter, initially intended by Mao as a means to purge his rivals in the party and purify the party ideologically, resulted in mass violence and a decade of political chaos. Among other results, the size of the police force shrank in the 1960s and then stagnated until the end of the Maoist era. Data from Zhejiang Province speak to the broader trend. In 1957 there were 7,502 provincial police; by 1962, the number had fallen to 6,002. The size of the force had rebounded to 10,500 in 1965, on the eve of the Cultural Revolution, but by 1978 had been reduced again to 6,268. Thereafter, the force recovered quickly, with an authorized strength of 23,800 in 1982.[17] Or consider Gansu. The province had 11,552 police in 1955; by 1964 their number had fallen to 6,165. Come 1982, the force had rebounded modestly to 7,860.[18] For its part, Guizhou Province had an authorized force of 11,272 in 1955, but budget cuts and purges resulted in a 30 percent reduction in force size by 1972. The provincial police did not regain their strength until after the Cultural Revolution.[19]

The magnitude of the twin shocks can be seen at the local level as well. In the 1950s, the number of police in Zhejiang's Xiangshan and Yin Counties averaged fifty-eight and fifty-four respectively. In the 1960s, the average number of police fell to forty-one in Xiangshan and thirty-four in Yin. In the 1970s, the size of the force recovered roughly to the level of the 1950s, but it had shrunk considerably relative to the size of the growing population. Then, in the 1980s, the force expanded, averaging 126 in Xiangshan and 156 in Yin. Finally, the 1990s saw explosive growth, likely due to the regime's decision to strengthen domestic security following the

Tiananmen crisis of 1989 and to growing fiscal resources. In Yin, the average number of police reached 332 in the 1990s.[20]

Spies and Informants

A top priority of the newly established MPS in the early 1950s was to develop a web of spies and informants. (In later chapters, I discuss the distinctions between spies and informants, as well as between various types of spies and informants.) As detailed by Michael Schoenhals, the MPS developed and operated an extensive network of *tebie qingbao renyuan* (special intelligence agents) prior to the Cultural Revolution. This was the beginning of a critical component of the surveillance state that operates to the present day.

The number of special intelligence agents is secret, but leaked documents and disclosures in *Major Events in Public Security since the Founding of the PRC* shed some light on the network of spies and informants established during the Maoist era, over which the MPS had exclusive control. Turnover was high, with agents serving an average of only a few years, and the total number of spies seems relatively small in the mid-1950s.[21] Minister of Public Security Luo Ruiqing disclosed in 1954 that the political security protection *xitong* (system) employed 23,000 spies, a number that he complained was too low. In addition, he was not impressed by the quality of his agents. Luo singled out Beijing, which had only 1,013 spies among a population of roughly 3 million.[22] Hubei province, with a population of nearly 30 million, had just 1,231 special agents in 1955, about 57 percent of whom worked for the political security protection unit, responsible for antisabotage and surveillance of political threats.[23]

The MPS had three categories of agents at the time. Case agents (*zhuan'an teqing*) gained access to groups and individuals under investigation. Intelligence agents (*qingbao teqing*) gathered information to help the police conduct secret investigations. Position agents (*zhengdi teqing*) protected critical areas and facilities and reported to police on suspicious activities occurring in spaces under their observation. Recruits for the first two categories were mostly from "enemy camps"—members of targeted groups who were usually coerced into spying for the police.[24] The MPS continues to use these three categories of spies today.

The spy network was devastated during the Cultural Revolution—as Schoenhals puts it, "Investigative work throughout the country was seriously damaged and a majority of agents were persecuted."[25] Indeed, the use of spies was suspended in December 1967, secret operating bases were dismantled, and the spies themselves were investigated. But the halt did not last long; spy operations resumed in the early 1970s. According to the MPS itself, the events leading to the resumption were remarkably petty. Apparently Premier Zhou Enlai became enraged when, in October 1972, Beijing residents dug up tens of thousands of flowers and plants on the south side of Tiananmen Square and took them home. Zhou blamed the success of this illicit act on poor intelligence and instructed that the MPS redeploy spies. In November 1973 the MPS formally resumed the use of agents nationwide.[26]

Local Security Committees

During the early Mao era, committees for public order and security *(zhi'an baowei weiyuanhui)* were the principal mass organizations performing auxiliary security functions, including surveillance. These committees, established in June 1952, were responsible for monitoring and "reforming" the "Four Category Elements" (*silei fengzi*)—landlords, "rich peasants," counterrevolutionaries, and bad elements—and for assisting the police in the effort of guanzhi, which was directed at counterrevolutionaries exclusively.[27] The committees worked under the supervision of local governments and public-security agencies. Each of the thousands of committees comprised three to eleven people, mainly volunteers drawn from the Communist Party and Communist Youth League.[28] Their composition speaks to the special capacity of Leninist regimes to deploy millions of party members and affiliated individuals to perform surveillance at low or even no cost.

Data from Jilin, Zhejiang, Fujian, Hubei, and Jiangxi during the years 1952–1954 show membership in these committees ranging from 0.31 percent of the provincial population to 1.15 percent. The average for the five provinces was 0.74 percent.[29] The regime's understaffed police service was thus augmented by an auxiliary force many times its size. Indeed, from the mid-1980s to the early 1990s, more than 12 million people served in these groups, representing about 1.1 percent of the population.[30]

The Political-Legal Small Group

In June 1958, the party created the precursor to the Central Political and Legal Affairs Commission (CPLC) that would become the main supervisor and coordinator of state surveillance in the post-Tiananmen era. This precursor, known as the Political-Legal Small Group, was headed by Politburo member Peng Zhen. The group included the heads of the Supreme People's Court, the Supreme People's Procuratorate, and the minister of the MPS. The group reported directly to the Politburo and the Secretariat of the Central Committee, of which Peng Zhen was also a member.[31] This marked the beginning of the practice of appointing the head of the body in charge of domestic security to the Secretariat of the Central Committee.

Although important as the precursor to the CPLC, the Political-Legal Small Group played only a minor role in coordinating domestic security policy during the Maoist era. Whereas the post-Tiananmen CPLC holds annual conferences setting the domestic security agenda, the Small Group convened only one such conference on record, in January 1959.[32] The first post-Mao conference did not occur until July 1982.[33] There were no real institutional equivalents of the Small Group at the local level. To be sure, most local party committees nominally had political-legal party groups (*zhengfa dangzu*), but local gazettes indicate that these had neither full-time staffs nor dedicated offices.[34]

Due to the passive role of the Small Group, the party relied exclusively on the national public-security conference (*Quanguo gong'an huiyi*) as the primary mechanism for policy coordination and implementation prior to the Cultural Revolution. Between 1950 and 1965, the CCP convened fourteen such conferences. Two were held during the Cultural Revolution (in 1971 and 1973), and five were convened between 1978 and 2019, reflecting the shift in the locus of authority to the CPLC.

Household Registration

As we have seen, the baojia was developed during the imperial period as a rudimentary form of household registration and as a law enforcement instrument. The baojia kept the state informed about the composition of families and was vital for the purposes of collecting taxes and levying corvee labor.[35] The Nationalist government relied on this system for social control and law enforcement as well, in the 1930s.[36] In the post-1949 period, the new registration system, known as the hukou, assumed a critical role in rationing scarce resources, controlling rural-urban migration, and enforcing the law.[37]

The hukou has also been essential for surveillance. The hukou required people to register their residential address and family members with the police, who conducted regular inspections to verify the recorded information. Households were also required to report updates, keeping police abreast of births, deaths, and relocations. The hukou became the institutional foundation of the Key Populations (*zhongdian renkou*) surveillance program, discussed below.

The hukou was potent but costly to build and maintain. Whereas the Mao era's underresourced police could outsource surveillance and control of class enemies to (mostly illiterate) volunteers and activists, tasks like registering births, updating household information, and approving relocations could be performed only by literate state agents. Accounts in local public-security gazettes indicate that insufficient police staff, coupled with the political shocks of the Great Leap Forward and the Cultural Revolution, so limited the development and effectiveness of the system that it could be enforced only in urban areas, where less than 18 percent of the population lived.[38] As a result, though the hukou was first implemented in 1951, it did not achieve its full effectiveness as a surveillance institution until the post-Mao era, when the party acquired adequate resources and technology to maintain a labor- and information-intensive system.[39]

The nationwide progress of the hukou was slow because of sparse police presence in the countryside. According to the MPS, the system was "preliminarily established" in more than half of county seats by late 1954, but this does not mean that many residents actually had been registered. In Jiangxi, the police reportedly completed household registrations in two-thirds of the rural townships—but not in the villages—by 1956, implying that most of the rural population was not covered.[40] In Hunan, only 10 percent of the population was registered by 1954. In 1959, 40 percent of Hunan's villages still had not set up the hukou.[41]

The January 1958 issuance of the Regulation on Household Registration marked the formal nationwide establishment of the hukou, with the police having exclusive authority over administration.[42] The government intended for the hukou to perform several important surveillance functions. Minister of Public Security Luo Ruiqing said that the hukou could "restrict the activities of, and sabotage by, counterrevolutionaries and other bad elements."[43] The police and local courts used the hukou to track individuals on parole, serving suspended sentences, or deprived of their political rights, who were required to obtain approval from the authorities before they could relocate to other jurisdictions. Counterrevolutionaries or other criminals might be discovered in the course of registering households. When Zhejiang implemented the regulation in 1958, the police used the hukou to identify targets for arrests, uncovering more than 1,500 counterrevolutionaries. In subsequent years, registration led to the discovery of previously unknown fugitives and other members of the Four Category Elements.[44]

However, even after full implementation, the hukou faced severe constraints. The lack of policing resources in the countryside forced the government to reassign administration to the people's communes—political and economic collectives established during the Great Leap Forward, which all rural residents were required to join—leaving their already-busy bookkeepers responsible.[45] More importantly, the Great Leap Forward arrived just after the hukou was fully established, degrading its effectiveness. The famine that resulted from the Leap led to a large migration of starving peasants to the cities, rendering the hukou unenforceable in many jurisdictions where police forces were too small to handle the influx.[46] Then, during the Cultural Revolution, a large number of police responsible for enforcing the hukou were either reassigned or discharged. Household-registration files were lost or destroyed. In many jurisdictions, the system effectively ceased to function. Revival of the hukou began in 1971, but it did not become fully functional until after the end of the Cultural Revolution, in 1976.[47]

Mass Surveillance Programs

The Maoist era saw three mass surveillance programs: guanzhi, the Four Category Elements (FCE), and the Key Populations (KP). As in the former Soviet Union, whose model of social control inspired the Maoist regime, individuals placed under these programs were deprived of many civil and political rights.[48] Because all of the surveillance programs were designed to restrict the activities of those considered political threats to the party, there was significant overlap among targets of the various programs. An individual labeled an FCE could be simultaneously placed under guanzhi and an individual under guanzhi was by default also a KP.

As we have seen, enforced control, or guanzhi, is a form of criminal punishment that restricts the activities of individuals designated counterrevolutionaries. Introduced in 1952, guanzhi was the province of local courts and public-security bureaus, which applied the punishment to those counterrevolutionaries whose past crimes did not warrant arrest and imprisonment. (Starting in November 1956, only local courts could sentence individuals to guanzhi.) Individuals under guanzhi were deprived of their civil liberties and political rights for up to three years at a time, and terms could be extended. After individuals were sentenced to guanzhi, the "masses" were notified of their sentences at public gatherings. Although nominally the police enforced guanzhi, in practice police relied on activists on local security committees to track the activities of those punished. During the Maoist era, guanzhi was the most formal and restrictive surveillance program in the country because it followed relatively clear procedures and it explicitly targeted counterrevolutionaries. Before 1959, the share of people under guanzhi was kept at under 0.15 percent of the population. After 1959, the maximum share was cut to 0.13 percent.[49]

Guanzhi ceased to function during the Cultural Revolution. In Jiangxi, police acknowledged that "the work of guanzhi descended into chaos"; in Tianjin, groups monitoring those under guanzhi were disbanded, subjects' files were destroyed, and the "work of mass supervision and reform was in a state of complete paralysis."[50] Guanzhi was revived in the post-Mao era, but after a revised criminal law went into effect in 1997, guanzhi was designated a penalty for petty crimes. In practice, it has now been merged into the KP program.

During the Maoist era, the FCE program counted the largest portion of the population under surveillance. Two of the four covered categories—landlords and rich peasants—were class-based. The other two—counterrevolutionaries and bad elements—were determined by political association or personal behavior. Counterrevolutionaries were further divided into "historical counterrevolutionaries" and "contemporary counterrevolutionaries." Historical counterrevolutionaries had past association with the Nationalist regime—the Kuomintang—or the Japanese puppet regime during the Sino-Japanese war (1937–1945). Leaders of banned sects also qualified as historical counterrevolutionaries. The label of contemporary counterrevolutionary was assigned to those who had no associations with the pre-1949 regimes but had committed counterrevolutionary acts in the wake of the CCP takeover. Finally, bad elements was a catchall category applicable to any undesirables who did not fit into the three other categories. Determination of who counted as counterrevolutionary and who as a bad element could be arbitrary. No national law or regulation specified exactly who belonged in which category, and there was no defined procedure for applying one label or another. Another category, rightist, was added in 1957 to cover those, mainly intellectuals, who were persecuted during that year's Anti-Rightist Campaign. In spite of the creation of a fifth category, the term "Four Category Elements" remained common. In January 1979, the FCE program was canceled, and the party removed the labels on landlords and rich peasants. It took five more years, however, before all labels had been removed.[51]

At various points, tens of millions of ordinary Chinese were designated FCE. The number of people classified at any given time was volatile, as the Maoist regime would push for more surveillance during its political campaigns and relent during less repressive periods. The central government did not actually do the labeling, nor did it collect data on the number of individuals falling under the various supervisory categories. Only local authorities—following often-vague diktats from the political center—labeled individuals, according to what were typically loose rules. And only local authorities collected information on the effects of the system.

TABLE 1.1
Share of Population Designated as Members of Four Category Elements, in Various Provinces

Province	Year	Share of population (%)
Guangdong	1956	1.7
Shaanxi	1958	0.82[a]
	1966	0.86
	1979	0.49
Tianjin	1973	0.54[a]
Jiangxi	1956	2.34
	1978	0.8
Hunan	1956	1.70[b]
	1973	0.65
Shanghai	1962	1.05
	1979	0.33
Fujian	1956	2.1
	1979	0.42
Gansu	1979	0.45
Jilin	1977	0.24
Zhejiang	1956	1.6
	1979	0.68
Guangxi	1979	0.60
Guizhou	1958	2.76[c]
	1960	2.36
Average	Pre-1977	1.46
Average	Post-1976	0.50

a Only those subject to "surveillance and reform" (*jiandu gaizao*).
b Only rural FCE admitted into communes.
c Estimate calculated on basis of population data for 1957 and 1961.
See notes for data sources.

The most authoritative disclosure of the scope of the program, contained in the *Chinese Law Yearbook 1987*, claims that more than 20 million people were labeled FCE in the years after 1949.[52] Estimating the share of the population labeled FCE at any given time is difficult because that total in a given year could be affected by deaths, new designations, and removal of the designation. Data from several provincial public-security gazettes for the Maoist period and the end of the 1970s provide rough estimates (Table 1.1): before the Cultural Revolution, an average of about 1.5 percent of the population, in any given year, was labeled FCE. The proportion fell dramatically, to around 0.5 percent, at the end of the 1970s, even before the party's decision to abolish the FCE program in 1979.

Available data suggest that FCE designations varied according to location: in rural areas, the FCE program chiefly targeted individuals designated through class labeling; in urban areas, more individuals were labeled on the basis of political activities and personal behavior. For example, in Shanghai, half of FCE in 1962 were counterrevolutionaries and bad elements, and the rest were landlords and rich peasants. By comparison, in more rural Zhejiang, landlords and rich peasants constituted 73 percent of FCE in 1959. Guizhou, a poor province, had a high share of landlords and rich peasants—82 percent—in 1960, while only 8 percent of FCE were labeled counterrevolutionaries.[53]

FCE were subject to varying degrees of surveillance. After establishment of the people's communes in 1958, rural FCE were divided into subcategories, with those under guanzhi facing especially onerous restrictions.[54] There are no national data on the share of FCE under guanzhi, but local data suggest wide variation. At the low end, in Zhejiang, only 3 percent of rural FCE were placed under guanzhi in 1956; in Hunan, only 5 percent of rural FCE were subjects of guanzhi at the time. At the high end, in Jiangxi, about 20 percent of rural FCE were under guanzhi that year, and in Fujian the share was 25 percent.[55] FCE not under guanzhi enjoyed greater freedoms, but they were nevertheless subject to constant surveillance by local security groups. In Guangxi, the names of FCE designated as subjects of "supervision and reform" (*jiandu gaizao*) were posted in public, as were the restrictions and rules they were required to obey. Shanghai adopted similar tactics so that the masses could monitor designees. As of 1962, about one-quarter of FCE in Shanghai were designated subjects of supervision and reform. In Guizhou in 1958, roughly one-quarter of FCE were under supervision and reform.[56]

As a general rule, each FCE would be under the watchful eyes of a "monitoring and reform team," which typically comprised ten "good people"—usually local officials, members of the local law and security committee, and other supporters of the party. The team monitored the physical labor and movements of the subject and regularly evaluated his or her performance; FCE were required to fulfill annual performance goals set by the team. At the end of each year, designees would be evaluated by the local community. Those deemed to have completed their reform would have their labels removed.[57]

The FCE program was imprecise, and its operations were mostly in the hands of volunteers—the masses. The regime had to make do because policing resources were limited, and there were lots of potential targets: the terror campaigns provided sufficient information about the population. But the result was that pro-regime activists operated a crude mass surveillance program that victimized huge numbers of people—more than 20 million over three decades. They suffered greatly. Many lost their freedom and dignity and were condemned to lives of misery, discrimination, and exploitation.

The Key Populations program, which targeted serious threats to regime security and public safety, was considerably smaller than the FCE program. While guanzhi and FCE were effectively outsourced to the masses, KP relied on the hukou to identify, register, and track designees, and only police had access to the hukou. KP thus posed severe administrative challenges to an understaffed public-security apparatus. Implementation difficulties were apparent in the late 1950s. Local PSBs in Zhejiang reported that the names of those targeted as KP were not updated for a lengthy period of time, most likely due to the disruption caused by the Great Leap Forward and austerity measures that resulted in cuts to the police force.[58] During the turmoil of the Cultural Revolution, KP was suspended.[59] Consequently, the program did not play a meaningful role in mass surveillance until the 1980s.

KP was instituted in March 1956 with the MPS's promulgation of the "Interim Rules on Managing Key Populations."[60] The targets were mainly historical counterrevolutionaries, although officially, there was a wide range of KP designations, overlapping those of the other programs. In 1962, the MPS included in the KP category "landlords, rich peasants, counterrevolutionary elements, bad elements, rightists, counterrevolutionary and other criminal suspects, and elements belonging to the social base of the counterrevolutionary class."[61] Counterrevolutionary elements were known counterrevolutionaries; counterrevolutionary suspects were under suspicion but not yet caught in the act; and the social base included family members of counterrevolutionaries as well as those, such as capitalists and landlords, whose occupation or socioeconomic status was associated with counterrevolutionary tendencies.

Disclosures about KP from several localities provide a glimpse into the scope of the program during the Maoist era. In 1959, Heilongjiang Province placed 27,325 individuals in the KP program, representing 0.16 percent of the population. Of these, 175 were suspected of counterrevolutionary or other criminal activities; 6,332 were common criminals and "hostile class elements"; and another 19,243 were listed as in need of investigation, control, or reform through education.[62] Presumably this last group included both ordinary criminals and political opponents. It appears that, in 1959, Heilongjiang prioritized a range of potential troublemakers; individuals suspected of counterrevolutionary activities constituted only a minuscule portion of the KP program. The lack of direct reference to historical counterrevolutionaries may indicate that these had already been imprisoned or executed during prior terror campaigns.

Data from several localities in the 1950s show that the program covered just a tiny percentage of the population. Only 0.06 percent of the population of Chongqing in 1958 was designated KP; on average, the same was true of six rural counties in Zhejiang. Hangzhou appears to be the only jurisdiction that placed a significant share of the population under KP: 0.59 percent during 1955–1958.[63] The dearth of data from the 1960s and the lack of disclosures in local public-security gazettes may indicate that, during the Maoist period, KP was more a concept than a substantive surveillance program, most likely due to the underdevelopment of the hukou, lack of police manpower, and constant political disruptions. In the post-Mao period, the apparently moribund KP program blossomed, strengthening sufficiently to become a pillar of the surveillance state.

Building the Surveillance State in the 1980s

The evolution of the surveillance state in the 1980s reflected the larger thrust of post-Mao China's development during the decade, which was a kind of interregnum between distinct political eras: the totalitarian Maoist era, marked by brutal repression, and the neo-authoritarian post-Tiananmen era, marked by pro-market economic development under a one-party regime. On the one hand, the coercive apparatus that had been severely damaged during the Cultural Revolution did begin to regain its capabilities. On the other hand, fiscal constraints still limited modernization of the surveillance state. And reformist leaders, who did not seek to move aggressively against political threats, were in charge of the party's day-to-day affairs.

The process of repairing the surveillance state began in 1982 with the party's new "Instructions on Strengthening Political-Legal Work." This order increased the size of the political-legal apparatus and emphasized covert operations against espionage and counter-revolutionaries.[64] This process of rebuilding was reflected in the growth of the police force, as noted above. According to the MPS, there were slightly more than 380,000 uniformed police in July 1972 (no information on the size of the force at the end of the Cultural Revolution is available).[65] By 1986, there were 600,000 uniformed police, an increase of 58 percent, with most of the growth likely coming in the early 1980s. In 1989, the MPS reported 769,000 uniformed police, for an average annual increase of

9.4 percent between 1986 and 1989.[66] Provincial public-security agencies grew, too. For instance, in Hubei, the number of authorized PSD personnel rose from 21,321 in 1979 to 33,374 in 1989, an average increase of 5.6 percent per year.[67]

The strengthening of the surveillance state was also a result of institutionalization and professionalization. The most important institutional development in the 1980s was undoubtedly the national ID card law enacted by the National People's Congress in September 1985. Eventually the issuance and upgrading of a unique national card would greatly facilitate state surveillance.

Another important institutional development was the establishment of the CPLC in January 1980. Peng Zhen, who had been a top party official responsible for domestic security in the 1950s, chaired the commission, which included the minister of public security and the heads of other law enforcement and judicial bureaucracies. The new commission's initial mission was largely confined to policy research and formulation.[68] Its ability to coordinate domestic security was limited by its small staff and the lack of manpower in local political-legal committees.[69]

The most substantive contribution of the newly established CPLC was the convening of national conferences, albeit irregularly, to set the domestic security agenda. The first national conference was held in July 1982. In 1983 the CPLC was also put in charge of Deng's "strike-hard" anti-crime drive, and in April 1983 it convened another key conference. Attendees recommended establishing a new Ministry of State Security and modernizing the domestic security apparatus with better technology, among other reforms. Party leadership approved all of the recommendations soon after.[70]

The Ministry of State Security (MSS) was established to achieve a more efficient division of labor between counterintelligence and domestic surveillance operations. In June 1984, party leaders approved a joint proposal by the MPS and MSS, titled "Opinions on Strengthening Cooperation between the MSS and the MPS." This unpublished document likely specifies the operational responsibilities of the two ministries.[71] Additionally, the MPS issued a series of new and updated rules on surveillance, such as its "Interim Rules on the Use of Informants" in 1978, operational rules on criminal investigations (1978), an opinion on strengthening intelligence collection (1979), a document on "strengthening basic-level work of urban police stations" (1980), and two revised versions of the KP-management regulations (1980 and 1985). Operationally, the MPS restored and expanded the KP surveillance program. In a December 1984 report, the ministry concluded that during the previous few years, the police had "strengthened the building of clandestine forces *(mimi liliang)*" a catchall term for spies and informants—"local security groups, and mass organizations" involved in security and crime-prevention.[72]

Throughout most of the 1980s, the MPS was firmly under the control of party hardliners.[73] The ministry was bent on strengthening its surveillance capabilities and operations against political threats. It convened regular meetings on political security, persistently emphasized the importance of building up its spying capabilities, carried out crackdowns on religious groups, and constantly demanded vigilance against, on the one hand, domestic counterrevolutionaries and "subversives" and, on the other, external hostile forces.[74]

Despite these improvements and the stridency of key institutions of the security apparatus, the surveillance state as a whole operated under serious resource constraints, which limited the party's ability to invest in technology and manpower. The government's total revenue growth from 1979 to 1989 averaged 8.3 percent per year, less than one-half of the growth rate of 18 percent per year between 1990 and 2012.[75]

Politically, the presence of reformist leaders such as Hu Yaobang and later Zhao Ziyang denied the coercive apparatus a clear mandate to adopt aggressive surveillance practices that could undercut the image of a China on a path to "reform and opening." Other senior party officials similarly stood in the way of a more empowered surveillance state. While they were not liberals, those in charge of the CPLC in the 1980s—such as Peng Zhen, Peng Chong, and Chen Pixian—had suffered terribly during the Cultural Revolution and apparently had no desire to revive Maoist totalitarianism.

When the coercive apparatus did receive clear political support, it was assigned tasks that diverted resources from surveillance of political threats—above all, Deng's 1983 anti-crime initiative, which resulted in excessive arrests, imprisonments, and executions.[76] By the time the campaign ended in January 1987, 1.77 million people had been arrested, and most were sent to prison. The exact number of executions is unknown; however, most scholars consider it likely that, if arrests were unwarranted, so were executions.[77] In addition to devoting scarce police resources to a campaign reminiscent of the worst excesses of the Maoist era, Deng's war on crime burdened the surveillance state with a large number of new targets. Since people released from prisons were automatically placed under the KP program, the police suddenly had to surveil a great many more people, some of whom may have been criminals but few of whom had ever been considered counterrevolutionaries or otherwise threatening to the regime.[78]

In retrospect, we may find here an explanation as to why the surveillance state failed to preempt the prodemocracy movements of the 1980s. Throughout the decade, intellectuals and college students advocating for democracy enjoyed unprecedented freedom in writing, publishing, networking, and otherwise pushing the boundaries for political change.[79] University campuses were hotbeds of liberalism, and there were major student-led pro-democracy movements in 1986 and 1989. A "culture fever," featuring open discussion of politically sensitive topics, struck in 1988. If the surveillance state did little to suppress these forces threatening the party's grip on power, it is likely because liberal reformers and hesitant conservatives denied the coercive apparatus the authority to conduct aggressive operations and because Deng's obsession with crime was a huge distraction.

The post- Tiananmen Surveillance State

After the party deployed troops and tanks to crush the Tiananmen pro-democracy movement in June 1989, it was determined to prevent a similar crisis from threatening its rule again.

The party also faced new dangers to its ability to maintain social order. Beginning in the early 1990s, China saw massive migration from the countryside to urban areas, as economic booms created more opportunities in cities and the government relaxed its control on internal migration. This severely strained the hukou. Greater labor mobility amid economic reform also loosened the ties that had bound urban residents to their state employers, entities that previously had been effective in monitoring and controlling their workers' behaviors.[80] In addition, because of increasing wealth and declining restrictions on information access and public speech, Chinese society was able to take advantage of novel communications technologies and was increasingly exposed to the outside world. Even if democracy movements had been squelched, civic space expanded by other means, creating more favorable conditions for political resistance and social unrest.[81] Disputes over land rights, wages, and pollution arose in the second half of the 1990s, testing a regime obsessed with stability.[82]

Indeed, the socioeconomic changes in the post-Tiananmen era challenged the surveillance state to an unprecedented degree. During the Maoist era, the targets of the surveillance state—primarily class enemies—were easier to define because of their political associations and socioeconomic status, and deprivations of their civil liberties simplified and facilitated surveillance. Additionally, restrictions on market-based economic activities, physical mobility, and access to information—which persisted through the 1980s, albeit to a lesser degree than in the Maoist period—meant that even a crude surveillance apparatus could, for the most part, meet the party's needs for social control. However, during the post-Tiananmen period, China's economic take-off engendered new sources of social and political conflict, placing greater demands on the surveillance state and encouraging party leaders to invest in more sophisticated mechanisms of coercion.

Thus, in the decades following the Tiananmen events, the government poured resources into modernizing and strengthening its surveillance of political threats and into maintaining social stability. Old methods of control were resurrected, refined, and updated; hardliners were appointed to oversee the security apparatus; and regime survival became the top priority of the party-state.

Regime Priorities and Resolve

In April 1990, in the immediate aftermath of the Tiananmen crackdown, the party issued a notice on "maintaining social stability and strengthening political-legal work."[83] This landmark document decreed that stability maintenance was the government's overriding mission. Substantively, the party instructed its local committees to prioritize political-legal work—party-speak for domestic security tasks—and provided the localities with necessary support in terms of staffing, funding, and political status. Another key document, the party's "Decision on Strengthening Public Security Work," arrived in October 1991. According to a leaked summary of the secret document, the party pledged to devote enormous resources to build up the capabilities of the coercive apparatus.[84]

In November, the MPS convened the eighteenth public-security conference—the first since December 1977.[85] Shortly thereafter, the agency launched several initiatives to prioritize surveillance of potential political threats, such as strengthening management of the KP program and bolstering security and covert operations on university campuses. In 1992 the MPS intensified "political security protection" by guarding against "infiltration and peaceful evolution"—a euphemism for Western pressure through educational, cultural, and commercial exchanges—and targeted underground Catholic groups and others considered "evil sects." The MPS's efforts did not slacken as the decade went by. In July 1997 it strengthened online surveillance, and in October it tightened control of foreign funding for Chinese social science research institutions.[86]

As a parallel initiative, in February 1991 the party leadership and State Council—essentially, the Chinese cabinet—issued a joint decision on "Strengthening Comprehensive Social and Public Order Management," which laid out a comprehensive strategy for safeguarding regime security and social stability.[87] (A follow-up order was issued in September 2001.) The key components of the strategy included full utilization of the capabilities of the public-security apparatus; strengthening local security outfits; mobilizing grass-roots security organizations; strict accountability to ensure that local officials implemented stability-maintenance measures; close cross-agency coordination under party leadership; equal emphasis on repression, prevention, education, and routine enforcement; and resolution and containment of social conflicts.

In the early 2000s, the party took additional steps to beef up its coercive capabilities. This new round of upgrading coincided with the elevation of hardliner Luo Gan as head of the CPLC and the appointment of a competent apparatchik, Zhou Yongkang, as minister of public security. Zhou would succeed Luo in 2007. As the heads of the CPLC from 1998 to 2012, the duo was instrumental in the modernization of the surveillance state.[88] Compared with measures taken by the party in the early 1990s, the efforts of the early 2000s were far more systematic and ambitious, as reflected in the landmark 2003 document, "The CCP Center's Decision on Further Strengthening and Improving Public Security Work."[89]

The document identifies a range of threats against which to guard: "infiltration" by external and internal hostile forces, ethnic separatists, sabotage activities by religious extremists and terrorist groups, illegal activities by the spiritual movement Falun Gong and other "evil sects," mass incidents, and individuals with the potential to form organizations that endanger state security and social stability. The party pledged to commit enormous resources to build up the grassroots operations of the public-security apparatus, fund public security reliably, expand the police force, increase police pay and benefits, and strengthen "policing through science and technology." Finally, the political status of local police chiefs was elevated to ensure the party's leadership in domestic security: the document orders local party committees to make local police chiefs members of the party standing committee and give them concurrent appointments as deputy governors or deputy mayors.[90]

Investment and Institutionalization

Thanks to the revenue increases produced by sustained economic growth, the party was able to plow immense resources into its coercive apparatus in the post-Tiananmen era.[91] In 1991, China's state revenue was 314.9 billion yuan; by 2020 it had reached 18,291 billion yuan—a twelvefold increase in real terms.[92] Table 1.2 shows the corresponding increase in domestic-security spending.[93] In nominal terms, spending on domestic security—excluding spending on the People's Armed Police—increased twenty-fourfold between 1991 and 2020, and about 1,900 percent after adjusting for inflation.[94]

The bulk of the spending was on the police, leading to rapid expansion.[95] In 1989, China's public-security apparatus consisted of 769,000 personnel; by 2010, the number had jumped to at least 2 million.[96] Data from Shaanxi, Hubei, and Zhejiang Provinces suggest that the police force grew at a much faster pace in the post-Tiananmen era than it did in the early to late 1980s. The number of authorized personnel in Hubei's provincial PSD rose from 33,374 in 1989 to 58,874 in 1999.[97] Zhejiang increased the size of its police force on nine separate occasions between 1990 and 2003, adding 7,840 positions, equivalent to 40 percent of its strength in 1985.[98] Guizhou added 10,674 police between 1991 and 2000.[99]

Local data show that the units responsible for domestic surveillance were expanded in 2000 and 2001. In 2000, the central government authorized Hubei to add 440 police positions solely for "domestic security basic work and flexible response units." In 2001,

TABLE 1.2
Spending on the Police, Procuratorate, and Courts, 1991–2020

Year	Amount (billions of yuan)	Share of total public spending[a]
1991	10	4.1
1995	30.5	6.19
2002	110	4.99
2004	154.8	5.43
2007	334	6.91
2011	522	4.78
2014[b]	702	4.62
2017[b]	10,467	5.15
2020[b]	11,645	4.74

a Refers to all spending by the central, provincial, and local governments.
b Figures for these years are estimates.
See notes for data sources.

the central government gave Hubei an additional 90 positions in domestic security units to combat "evil sects."[100] Zhejiang received 540 authorized positions for its domestic security unit in 2001 as well. While it is only possible to access information from individual provinces, most likely police units responsible for domestic security were authorized to grow nationwide, and probably they added more than 10,000 agents.[101]

While the state invested in manpower, it also upgraded its technological capacity for coercion. In November 1991, the MPS convened a national conference on the modernization of law enforcement technology, where it announced that the country's broader technological modernization program would incorporate the needs of "science and technology for public security."[102] In August of the following year, the MPS began to build a national crime information data center, taking the first step toward digitizing its information system. Soon after the MPS issued its May 1998 "Decision on Strengthening Science and Technology Work in Public Security," the party formally approved the Golden Shield project, an IT modernization program that included the so-called Great Firewall of China.[103] In 2004, the MPS launched Skynet, a high-tech video and sensor surveillance program. Eleven years later, in 2015, the CPLC rolled out Sharp Eyes, another video surveillance program that complements Skynet.

Simultaneously, the party reengineered the institutions of the surveillance state, beefing up the CPLC and its local outfits. These enhancements took multiple forms. One involved signaling the priority granted to surveillance by elevating the political stature of security institutions. For instance, the party elevated the political status of the local political-legal committees by directing that the party official in charge also be made a member of the standing committee of the local Communist Party committee or a deputy secretary of the same local party committee. Similarly, a new commission, the specialized Central Commission on Comprehensive Social and Public Order Management, was created within the CPLC. This was a shell commission; it had no staff and no tasks. But within the traditions of Chinese politics, its existence testified to the importance of the work the CPLC was doing. More substantively, local political-legal committees were granted a key role in the appointment and promotion of law enforcement officials. The revived CPLC and local political-legal committees received a broad mandate to provide "macro-level guidance" and coordinate the work of the domestic security sector.[104] And the CPLC and its local affiliates established annual national political-legal work conferences—and their local equivalents—starting in 1990. Whereas the CPLC in the 1980s convened such conferences irregularly and infrequently, since 1990 the CPLC and its local political-legal committees have held annual work conferences without fail.

In addition to strengthening the CPLC and the local political-legal committees, the party established more specialized offices to coordinate responses to specific political threats. These include "stability-maintenance offices," established in the mid-2000s and tasked with addressing social conflicts, such as strikes, protests, and riots. Formally part of the CPLC and the local political-legal committees, stability-maintenance offices have full-time staffs and operate their own networks of informants.[105] To suppress Falun Gong and other spiritual organizations, the party set up special offices for combating "evil cults," commonly known as the 610 Offices, which were again under the umbrella of the CPLC and the local political-legal committees. (The 610 Offices were abolished in 2018.) In August 1998, the Public Information Network Security Bureau, a special police unit that monitors the internet and combats cybercrimes, was formed inside the MPS and its corresponding local outfits—another effort to address a novel threat.

The Xi Jinping Era

By the time Xi Jinping rose to power in late 2012, he had at his disposal a fully modernized surveillance state. Yet, despite the revival of totalitarian practices under Xi, development of the surveillance state has been marked by contradictory trends during his rule. On the one hand, Xi has continued the technological upgrading of surveillance, in particular strengthening cybersurveillance. He established the Cyberspace Administration in 2014 to coordinate control of the internet; in 2018, the administration was elevated to a central commission, granting it broad authority.[106] To further facilitate state surveillance, his government passed a cybersecurity law in 2016 and a data security law in 2021. As noted above, the high-tech Sharp Eyes project was launched in 2015, under Xi's watch, and Skynet has been upgraded. Xi's government can take credit for full implementation of the grid management system and the launch of the social credit scheme, a data-driven system with which authorities may be able to track individuals' behavioral patterns and even political loyalties.[107]

On the other hand, Xi's government has faced real fiscal constraints, as economic growth decelerated after 2012. Consequently, state revenue averaged an anemic 5 percent growth per year between 2013 and 2020, compared with 20 percent between 2003 and 2012.[108] This led to slower increases in domestic security spending. Between 2013 and 2020, average nominal growth in spending on domestic security was 12 percent per year, compared with an astonishing average nominal growth of 35 percent per year in 2003–2012.[109]

Xi has also taken aim at elements of the domestic security apparatus to ensure its political loyalty. Shortly after he assumed power, he purged Zhou Yongkang, the former head of the CPLC (2007–2012). The political status of the CPLC was also downgraded: its head was allowed to remain a member of the Politburo but not of its standing committee. The formal authority of the CPLC appeared to be undermined further in 2014 when the party established the Central National Security Commission, with Xi as its chairman.[110] Starting in 2018, Xi launched a three-year campaign to purge the domestic security apparatus, resulting in the arrest and punishment of tens of thousands of police officers, including four former vice ministers of the MPS, many provincial police chiefs, heads of the provincial and local political-legal committees, and large numbers of local police chiefs, prosecutors, and senior court officials.[111] As I discuss in more detail later, these changes do not constitute a downgrading in the importance of domestic security. Rather, this is Xi's effort to address the coercive dilemma: he has filled open positions with officials and police officers loyal to himself.

The Surveillance State during the COVID Pandemic

The capabilities of China's surveillance state were fully mobilized during the COVID-19 pandemic that began in Wuhan in December 2019. Immediately after the outbreak, the Chinese government deployed its surveillance capabilities to enforce the so-called zero-COVID measures, intended to eradicate the virus. Although not designed as a public-health tool, China's surveillance state was readily repurposed because it possesses the technological and organizational means to track ordinary people's movement and activities, which is useful in locating sick people and subjecting them to quarantines.[112]

On the technology side of the ledger, the party relied almost exclusively on smartphone-based tools—such as apps and location monitoring via GPS—to track the health status, social contacts, and movement of individuals.[113] Specifically, the government collaborated with private technology companies such as Alibaba and Tencent and with state-owned telecommunications firms like China Mobile and China Unicom to develop and implement a health code (*jiankangma*) and a travel code (*xingchengma*).

In theory, the health code relied on user input and big data analytics to assess individuals' health status and assign them to one of three color-coded groups: red, yellow, and green. A user with a green code could move freely, accessing public places such as shops, restaurants, schools, and hospitals. Those with a red or yellow code—the former denoting high-risk individuals and those diagnosed with COVID, and the latter indicating moderate risk—would be forced to quarantine at home. The system was based on an Alibaba smartphone app deployed in February 2020 to track the health status and close social contacts of the company's employees. The municipal government of Hangzhou, where Alibaba is based, quickly embraced the app. Shortly afterward, all provincial governments partnered with technology companies (primarily Alibaba and Tencent) to roll out their own schemes; there was never a single national system. The code was generated by apps linked to WeChat, an app developed by Tencent and used by more than a billion people in China, and to Alipay, a digital wallet developed by Alibaba and widely used in the country. WeChat and Alipay users were required to enter their personal information (national ID number, age, gender, and address) and health-related information (vaccination status and results of COVID tests). The movement of the users was tracked so that the WeChat and Alipay apps would know if they came into close contact with an infected person.

Exactly how the health code actually operates has never been clear. The algorithms behind the apps remained opaque, and, again, there were no national standards. Based on press reports that local authorities misused the health code to restrict the movement of protesters and dissidents, it is clear that local authorities could use the systems for purposes unrelated to tracking potential spreaders of disease.[114] Before the rush to abandon zero-COVID in December 2022, the health code seriously circumscribed the daily lives of ordinary people, including those without COVID. Even those with green codes could hardly live normal lives, for fear that a chance encounter would result in a change of status to yellow or red, regardless of their own health condition.

Besides the health code, China also introduced an app to track individuals' travel history. Drawing on location data provided by the three state-owned telecom companies, the app generated a log of locations to which an individual had traveled in the previous two weeks. Authorities could use the log to ascertain whether an individual had traveled to outbreak areas or had been in contact with infected passengers on public transportation. On December 11, 2022, in the wake of the fierce protests that brought the zero-COVID policy to an end, the government abolished the travel code and claimed to have deleted all the related data.

Although the health and travel codes likely did serve their official, publicly announced purpose of protecting public health, they also enabled major advances in the party's use of surveillance technology on a population-wide scale. For the first time in Chinese history, the state was able to use advanced technologies to determine the health status and travel history of ordinary people and restrict their freedom of movement accordingly. The heaps of health and mobility data collected during the pandemic might also be a valuable resource in enhancing the state's surveillance capabilities in the future.

Indeed, the potency of the health code as a dual-use tool was such that, in November 2022, the Chinese government announced a plan to build within three years a unified national health platform collecting digitized health records of all citizens.[115] Once completed, this platform could generate certain identifying markers, such as the colored health codes now familiar to ordinary Chinese, that would further strengthen the state's capacity to monitor and control citizens.

It would be a mistake, however, to treat the zero-COVID measures as purely technological efforts. The state also deployed grid management, a labor-intensive instrument, to enforce zero-COVID with great effectiveness. Grid management, the institutional origin of which I previously traced to the baojia, was introduced in Beijing in 2003 to prevent crime and improve the operations of urban facilities, but it gradually evolved into a powerful tool of state surveillance.[116] This system divides communities into several grids, each of which typically contains about 300 households. An attendant, who is either a part-time or full-time employee of the local neighborhood or village committee, is assigned to each grid to assist law enforcement and provide real-time reports on infrastructure problems and minor incidents, such as traffic violations. Authorities in wealthier communities have equipped their grids with advanced technologies, such as specialized apps and digitized information concerning public facilities. Even before the pandemic, grid management had been used to surveil targeted individuals.

The pandemic presented the first opportunity to use grid management to enforce a central-government policy nationwide, as the system complemented the health and travel codes in maintaining zero-COVID. At a meeting of the Politburo Standing Committee during the outset of the pandemic in early February 2020, Xi instructed the party to "strengthen grid management in communities" to contain the outbreak in Hubei.[117] This made sense because vaccination drives, mass testing, disinfection, verification of the health code, enforcement of home-based quarantines, and imposition of lockdowns at larger-scale facilities (such as hotels and apartment buildings) all require labor-intensive efforts at the community level. The existing system of grid management provided a structure for deploying those efforts.

Following Xi's instruction, local governments in Hubei and beyond relied on grid management to report infections and travelers from infected areas, provide essential services (such as delivering groceries to families under lockdown), staff access points to apartment complexes, and communicate official announcements. Attendants sent residents of their grids messages via WeChat, conducted disinfections, checked health codes, and assisted in mass COVID testing and promotion of vaccination.[118]

The wide-ranging responsibilities performed by grid management during the pandemic put the operational capabilities of the system to the test. As a result, this mechanism of social control has likely been significantly improved. At a minimum, grid attendants empowered to enforce public health measures likely acquired new skills and gained an unprecedented amount of information about ordinary people. Then, too, authorities probably acquired valuable knowledge concerning the functioning of the system under stressful conditions and over an extended period of time.

DURING THE MAOIST PERIOD, the regime established the institutional and organizational foundations of the surveillance state: a centralized public security bureaucracy, a network of spies, volunteer security organizations, mass surveillance programs, and the household-registration system. The framework of distributed surveillance thus was firmly in place during the first two and half decades following the revolution. The post-Maoist surveillance state, which is the focus of the remainder of this book, rests on this foundation. To offset its lack of material resources, the Maoist regime maximized its brutality and organizational capabilities. The party took advantage of popular support and of the ideological commitment of its members to maintain a vast, labor-intensive surveillance apparatus at little financial cost. The "mass line"—mobilization of the general public to perform routine, low-level security tasks—was adopted and promoted widely. This approach to surveillance has received the endorsement of Xi Jinping, and it remains a crucial instrument of preventive repression today.[119]

The surveillance state that was developed during the Maoist period nevertheless had serious weaknesses. The lack of resources, as well as the regime's inability to develop more sophisticated institutions of command and coordination or to acquire advanced technologies, limited its reach. Ultimately, the Maoist government turned out to be its own worst enemy. Its ideological extremism, which inspired the Great Leap Forward and the Cultural Revolution, severely disrupted and degraded the surveillance state that was meant to protect that very same government.

The story of China's surveillance state in the post-Tiananmen era could not be more different. Relying on the institutional foundations of the Maoist totalitarian regime, the party has significantly upgraded the surveillance state through investment and institutionalization. Relative political stability, and the regime's decision to prioritize security and social stability, enabled the surveillance state to gain new operational capabilities and tactical sophistication. The adoption of advanced surveillance technologies during and after the late 1990s has added a new dimension to distributed surveillance, enabling the party more effectively to perform certain tasks, such as monitoring digital communications and movements of targeted individuals. What makes the contemporary Chinese surveillance state uniquely formidable are its human resources and Leninist hierarchical structure, coupled with ample funding and advanced technologies.

CHAPTER 2

Command, Control, and Coordination

All dictatorships face two major challenges in prosecuting effective state surveillance. The political challenge lies in addressing the coercive dilemma: to maintain the loyalty of officials in charge of the repressive apparatus, in particular the secret police. Ambitious and disloyal individuals in these pivotal positions have the means to spy on their masters and conspire with their masters' rivals.[1] The second challenge is an operational one: to ensure that the regime's security agenda is fully implemented. The institutional arrangements and organizational capacities of dictatorships determine their ability to address these two challenges.

In hopes of solving the coercive dilemma, rulers in personal dictatorships typically rely on trusted loyalists to run their secret police. For example, the longtime director of Iran's SAVAK, Nematollah Nassiri, was a close friend of the shah. For his part, Saddam Hussein hired his second cousin to direct the intelligence service, and like Saddam himself, many of the dictatorship's officials hailed from the area of Tikrit.[2] Meanwhile, in military juntas, the generals in charge usually appoint one of their own to head the secret police, as was the case in Chile.[3] Finally, one-party regimes typically place the secret police in the portfolio of a senior party official. In the former Soviet Union and its satellites, a member of the Politburo usually served as chief of the secret police. In the case of East Germany, the ruling party set up a specialized party organization, the Central Committee Department for Security Questions, to supervise the work of the Stasi.[4]

The challenge of ensuring effective policy implementation and operational coordination among the several components of the coercive apparatus is arguably more daunting than even the coercive dilemma. The bureaucratic agendas and interests of the coercive apparatus often diverge from those of the rulers.[5] The coercive apparatus may not treat the ruler's top-priority tasks with urgency. Rivalries among bureaucracies may impede information-sharing and cooperation. Local authorities and national bureaucracies may struggle to work together because they are under different chains of command.[6]

Few dictatorships have developed institutionalized mechanisms, such as regular conferences or specialized organizations, to supervise implementation of the ruler's security agenda and to coordinate activities among the agencies of the coercive apparatus. Politics is obviously one explanation. To achieve an adequate level of coordination requires, at a minimum, a politically powerful organization that embodies the authority of the regime. This supervising entity is likely to be resented, if not resisted, by the rest of the security apparatus, whose individual department leaders fiercely guard their power and autonomy. Another practical explanation is cost. Besides being politically empowered, a specialized bureaucracy must operate at every level of the state in order to provide effective coordination. It therefore requires large numbers of personnel and facilities—an expense only dictatorships with plentiful fiscal resources can afford. Even the Department for Security Questions of the Central Committee in East Germany had no local branches.

China, too, struggled with implementation and coordination in the Maoist era. Although the CCP was able to establish the basic institutional framework of distributed surveillance, resource scarcity prevented the party from building and maintaining a large formal coercive apparatus, much less a costly stand-alone bureaucracy to coordinate repression. Except for national conferences on public security, the Maoist regime had no institutionalized mechanisms for coordinating repression and surveillance.

The party did not significantly improve its coordination capacity in the 1980s. The newly formed Central Political and Legal Affairs Commission (CPLC) and its local affiliates had small staffs and limited mandates. Only in the post-1989 era did the party succeed, albeit incrementally, in ensuring both the political loyalty and the operational effectiveness of the surveillance state. Its most important institutional innovation during this period was the strengthening of the CPLC and the establishment of fully staffed local political-legal committees. Between them, the CPLC and the local committees supervised and coordinated the activities of the coercive apparatus and other parts of the party-state.

In this chapter, I begin by examining the variety of mechanisms that today's Chinese party-state uses to coordinate coercive operations. It makes for a truly dizzying array. I then turn to the key institution, the one that oversees and organizes the others: the CPLC.

Commissions, Leading Small Groups, and Conferences

As in other Leninist party-states, the making of domestic security policy in China is centralized at the top. The Politburo Standing Committee (PSC)—currently comprising the seven top leaders of the CCP, with Xi Jinping occupying the first seat—gives initial approval of major domestic security initiatives and then charges party and state bureaucracies with implementing them. Policy initiatives may be proposed by the PSC members themselves or by functional bureaucracies, such as the Ministry of Public Security, or by the CPLC.

The party itself serves an important coordinating function through its assortment of conferences, meetings, commissions, and small groups. At the national level, one finds central commissions and leading small groups (LSGs). These consist of the heads of the various departments of the CCP, the ministries of the State Council, and other entities such as the judiciary and the military. The members of central commissions and LSGs meet infrequently and likely play a largely pro forma role in making and coordinating policy. The routine—and more substantive—functions of the central commissions and LSGs are performed by fully staffed offices, typically headed by the vice minister of a ministry with special competence in the relevant policy area. To avoid duplication of effort and resources, the party prefers to create commissions within an existing bureaucracy, a practice known as "two organizational titles with the same staff."

Central Commissions

The central commissions propose policies and supervise and coordinate implementation. Their members include heads of key party departments and government ministries who have responsibilities or expertise in the relevant policy area. The number of such commissions has increased since Xi Jinping became party chief in late 2012. The most notable central commissions established under Xi are the National Security Commission, the Cyber Security and Informatization Commission, the Foreign Affairs Commission, and the Central Commission for Integrated Military and Civilian Development.[7] These commissions meet infrequently; their day-to-day operations are performed by a special office attached to each. These offices have their own high-ranking leadership. Underscoring the special status of the National Security Commission, the director of its office is Xi's chief of staff.

In recent years, domestic security policy has been the purview of the CPLC, albeit with specialized involvement by the Cyber Security and Informatization Commission. Between 1991 and 2018, another central commission tasked with domestic security policy, the Central Social and Public Order Comprehensive Management Commission (*Zhongyang shehui zhi'an zonghe zhili weiyuanhui*), or SPOCMC, operated within the CPLC on the "two organizational titles with the same staff" model. The head of the CPLC usually concurrently served as the head of the SPOCMC.[8] No new bureaucracy was created, and the SPOCMC and CPLC were essentially the same organization. At the subnational levels, provinces, cities, and counties adopted the same model. Nominally, they all had their own committees of comprehensive management, but these committees were all located inside the political-legal committees, and they shared the same personnel.

Thus, if the SPOCMC was nominally the CPLC's equal in the state hierarchy, the CPLC and the party's local political-legal committees remained the center of the action. Between them, they oversaw, and now continue to oversee, implementation of domestic security policy. They coordinate the operations of the surveillance state. Notably, unlike other central commissions that rely on attached offices, the CPLC is itself a fully staffed commission. It is in this way it is similar to CCP top-level departments.

Leading Small Groups

Leading small groups are less formal and prestigious than central commissions. Organized to formulate national policy and ensure its implementation, LSGs can be thought of as high-level task forces.[9] Members of these groups, like commission members, are heads of key party departments and government ministries. LSGs meet at irregular intervals, and their routine functions are performed by full-time staffers working within ministries or departments with jurisdiction over a particular policy area. LSGs can have a number of areas of expertise, including security. Until their abolition in March 2018, the Central Leading Small Group on Safeguarding Against and Handling the Problem of Evil Cults and the Central Stability Maintenance Leading Small Group made policies concerning the suppression of banned spiritual groups and of social unrest, and these groups worked to ensure policy implementation nationwide.[10]

The Evil Cults LSG was established on June 10, 1999, under a different name. Its operational responsibilities were assigned to a special office that we have already encountered—the so-called 610 Office, which was located within the CPLC and directed by a vice minister of public security.[11] At the subnational level, the 610 Office was located inside local political-legal committees. Before its abolition in 2018, when its responsibilities were split between the CPLC and the MPS, the 610 Office played a critical role in the party's campaign against Falun Gong and other spiritual groups the party-state deemed undesirable.[12]

The importance of the Central Stability Maintenance LSG, established in 2000, was demonstrated by its leadership: it was chaired by the head of the CPLC. Its office, directed by the executive vice minister of public security, was located inside the MPS.[13] Few operational details about this office are disclosed in official media, except for some information about its research into causes of social instability. For example, public reports indicate that in 2014 the office sent a delegation to a Shandong state-owned steel mill to study how reducing "excess capacity"—that is, closing factories—would affect social stability. We also know that the deputy director of the office led two delegations, in 2009 and 2015, to conduct research in Jiangxi and Shandong.[14] Given what is known about the typical role of LSGs in policymaking and coordination, the Central Stability Maintenance LSG probably developed domestic surveillance proposals for consideration by the PSC.

Coordination Small Groups

Another institutional form underlying China's security state is the coordination small group. As indicated by the name, these groups do not participate in policymaking; rather, they coordinate the implementation of policies made by bodies such as the PSC. There are three coordination small groups engaged in domestic security, one concerned with domestic security generally and one each focused, respectively, on Tibet and Xinjiang.

The coordination small groups for Tibet and Xinjiang are chaired by the PSC member in charge of ethnic-minority affairs. Again, the fact that a PSC member—one of the seven top figures in the Chinese leadership—is at the helm indicates the importance granted to the group. Each of the two coordination small groups has a fully staffed office. According to one report, the office of the Xinjiang Work Coordination Small Group was initially set up inside the CPLC in 2000, but it was relocated to the State Ethnic Affairs Commission in 2013. A more recent report discloses that in 2019 the director of the office was a deputy director of the CCP's United Front Department, whose mission is to seek allies and support from social elites and groups that the party sees as strategically important, such as prominent religious figures, overseas Chinese, and ethnic minority leaders.[15] While the Xinjiang group focuses exclusively on that province and its Uighur population, the Tibet group is concerned with Tibet and four other provinces with large Tibetan populations: Sichuan, Qinghai, Gansu, and Yunnan and is led by the same PSC member who heads the Xinjiang group. Whereas the deputy head of the coordination small group for Xinjiang is the region's party chief, the deputy head of the coordination small group for Tibet is the director of the United Front Department. We have no information about the bureaucracy to which the Tibet group is attached, but its likely institutional home is also the State Ethnic Affairs Commission.[16]

The general domestic security coordination group is called the Safe China Construction Coordination Small Group. It is apparently the successor to a commission, the Central Social and Public Order Comprehensive Management Commission. Established in 2020, the Safe China group is led by the chief of the CPLC; other members of the CPLC attended the Safe China group's first meeting, indicating that they are also members of the group. Although technically its role is to coordinate domestic security policy implementation across departments, the Safe China group is very likely involved in making policy proposals as well. After all, its leaders include CPLC members: high-ranking national-level politicians.[17] The Safe China group has four specialized teams: the Law and Public Order in Society team (*Shehui zhi'an ju*), headed by a deputy minister of public security; the Public Safety team, headed by a deputy minister of emergency management; the Urban Area Management team, headed by a deputy secretary-general of the CPLC; and the Political Security team, headed by another deputy secretary-general of the CPLC.[18] It is probable that this latter group has the primary mission of coordinating surveillance and political repression.

Conferences and Meetings

At regular work conferences and meetings, officials from central- and subnational-government departments gather to mobilize the coercive apparatus. One of the critical functions of such conferences is to communicate the top leadership's security agenda outward to the provincial and local levels. Among these gatherings, the most important and prestigious is the National Public Security Conference. Of the twenty-one conferences convened since 1949, only five have been held in the post-Mao era, perhaps because the annual work conferences of the CPLC have largely supplanted the national public security conference. As of this writing, the last such conference was held in May 2019. Typically, the conference is preceded by issuance of a CCP document laying out a new medium-term domestic security agenda. For example, on November 18, 2003, the CCP issued its "Decision on Further Strengthening and Improving Public Security Work," just before the convening of the twentieth National Public Security Conference.[19] The presence of the top leaders of the party-state at these conferences is indicative of their status and importance. In the post-Tiananmen era, all top leaders (except Hu Jintao in 2003) met the conference delegates in person.[20]

If the National Public Security Conference is designed to mobilize the coercive apparatus (mainly the MPS) to implement the regime's medium-term domestic security agenda, annual domestic security priorities are communicated to the party-state through the CPLC's Political-Legal Work Conference (*Zhengfa gongzuo huiyi*). At the national level, this conference is held at the end or the beginning of each year. Provincial conferences are convened soon thereafter, followed by municipal and county-level conferences.

Separately, the MPS convenes its own meetings to implement the agenda laid out at the annual political-legal work conferences. At the national level, the annual MPS conference, held at the beginning of the year after conclusion of the CPLC's work conference, is called the National Police Chiefs Conference (*Quanguo gong'an juzhang huyi*). At the subnational level, provincial political-legal work conferences take place shortly after national conferences, and municipal conferences are convened after the provincial conferences.

This top-down mechanism of bureaucratic mobilization and policy implementation is a classic feature of a Leninist party-state. But even a well-formed hierarchy cannot ensure that local governments and security agencies will in fact faithfully turn the edicts of top party leadership into action, nor that they will do so effectively. The leading small groups and specialized offices have no corresponding local equivalents, while work conferences lasting a mere few days accomplish little beyond information-sharing and short-term bureaucratic mobilization.

The Central Political and Legal Affairs Commission

In all likelihood, the party found its answers to the challenges of coordinating the coercive apparatus—and maintaining its political loyalty—through trial and error. As we have seen, the party's late-1950s LSG on political-legal work played only a minor role. In the 1980s, the party elevated the status of the group to a commission, creating the CPLC. Yet the CPLC initially had a weak organizational presence at the local level.[21] In terms of domestic security policy formulation and implementation, it was not nearly as active as it would become after 1989.

Post-Tiananmen investments brought about the full-fledged institutional development of the CPLC and its local outfits. During this period, the party gradually built a vertically integrated institution with expanded supervisory authority over the coercive apparatus and sufficient organizational resources to implement domestic security policy. As a fully staffed CCP bureaucracy present at all levels of the party-state, the CPLC was placed on equal footing with the other four major CCP bureaucracies having their own corresponding departments at each level of the party-state: the Organization Department (responsible for vetting millions of officials), the Propaganda Department (which spreads official ideology and controls the media), the Central Commission for Discipline Inspection (China's top anticorruption body), and the United Front Work Department (which manages relations with nonparty individuals and organizations at home and abroad). As these departments are institutional pillars of one-party rule, the establishment of a security-focused party bureaucracy that replicates their nationwide hierarchical organizational structure signals the priority the party attaches to regime security. In addition, the decision to follow the model of the four pillars demonstrates the party's faith in the Leninist system. The existing organizational structures for mobilization of resources and implementation of policy were working, so the CCP applied them to the repressive apparatus it designed to protect itself.[22]

The CPLC, then, became the paramount overseer of domestic security, on the Leninist hierarchical model. With the aid of its subordinate bureaucracies and advisory groups, discussed above, the CPLC is responsible for translating the party's orders on domestic security into specific policy measures, coordinating actions among security agencies, supervising the work of the courts and procuratorates, and overseeing implementation of high-priority tasks such as law-and-order initiatives, crackdowns on dissent, and the buildout of high-tech surveillance systems. Based on a brief disclosure in official MPS documents, we know that the CPLC reports directly to the PSC, technically through the Secretariat of the Central Committee, on which the head of the CPLC sits. The CPLC receives its policy instructions from the PSC and submits domestic security reports to that same body.[23]

Today the CPLC is led by the Politburo member responsible for domestic security. (This actually represents a demotion: in the decade prior to Xi's rise, the head of the CPLC was one of the members of the PSC, which makes day-to-day decisions for the twenty-five-member Politburo.) His executive deputy is usually the minister of public security, signifying the critical importance of the Ministry of Public Security in domestic security. But the person wielding substantive power over the operations of the CPLC is its secretary general, equivalent to a Western minister's chief of staff. Other members of the CPLC include the heads of the Supreme People's Court and the Supreme People's Procuratorate, the minister of justice, and the commander of the People's Armed Police. Top national security and intelligence officials are also members, including the heads of the political-legal commission of the People's Liberation Army (PLA) and of the Ministry of State Security, a secret police agency combining elements of domestic and foreign intelligence and counterintelligence.

To prevent the coercive apparatus from acquiring too much power, the party imposes term limits on the head of the CPLC. Typically leaders serve one five-year term, although Luo Gan served two terms, from 1997 to 2007. At the subnational levels, most political-legal committee heads serve one five-year term, and the secretaries of the provincial committees are often rotated to prevent them from establishing fiefdoms.

Evolution of the CPLC since 1989

The political-legal sector encompasses a range of bureaucracies. In addition to the courts and the procuratorate, this sector includes the MPS and the local public security bureaus (PSBs), the Ministry of State Security and its local agencies, the People's Armed Police, the Ministry of Justice (which also oversees the prisons) and its local agencies, and the police academies. Since the founding of the PRC in 1949, the CCP has maintained a tight grip over the political-legal sector through specialized committees and their permutations, such as the Central Political and Legal Affairs Small Group that we have already encountered. After the formation of that central group in 1958, provincial, prefectural/municipal, and county-level CCP committees all set up political and legal affairs small groups and empowered them to coordinate the work of law-enforcement agencies.[24]

However, these groups were restricted to involvement in criminal cases and legal disputes. They had no role in formulating, implementing, or coordinating social control and surveillance measures. At the local levels, political-legal small groups did not have fully staffed offices. As the chaos of the Cultural Revolution engulfed the country, the political and legal affairs small groups ceased to function. Shortly after the Cultural Revolution, the CCP reestablished the Central Political and Legal Affairs Small Group and gave it a limited mandate to conduct research on major policy issues and assist the work of the Supreme People's Court, the Supreme People's Procuratorate, and the MPS.[25]

When, in 1980, the Central Political and Legal Affairs Small Group was elevated and transformed into the CPLC, the central and subnational political and legal committees together became a functional department of the party. Their responsibilities included "maintaining communications and providing guidance to the various departments in the political and legal sector; assisting party committees and bureaucratic departments in evaluating and managing cadres; organizing and conducting research on policy, law, and theory; coordinating joint meetings to deal with major and difficult cases; and organizing and promoting implementation of "comprehensive social management."[26] Despite their expanded mandate, the CPLC and its subnational units were not directly involved in routine operations or decisions of the law-enforcement apparatus. The reference to implementing unspecified comprehensive management, however, foreshadowed the pivotal role of the CPLC and the PLCs in supervising the surveillance state more than a decade later.

In between the emergence of the CPLC and its full-scale expansion in the 1990s, it was briefly abolished. In May 1988, under the leadership of CCP General Secretary Zhao Ziyang, reformers seeking to separate the party from the state eliminated the CPLC and replaced it with a new central political and legal affairs LSG with limited power. But the Tiananmen crackdown came only a year later, and there was not enough time to fully implement the reform at the subnational level, leaving most provincial and local PLCs intact. In 1990, soon after the crackdown, the party reestablished the CPLC.[27]

As defined during the reestablishment of the CPLC, the principal mission of the political-legal committees is to provide "macro-level guidance and coordination" and to advise and assist party committees. PLCs are subordinate to the party committees at the same administrative level, but they receive guidance from their superior PLCs. The specific responsibilities of the local PLCs are determined by party committees. The 1990 document reestablishing the CPLC and subnational PLCs elevates the political status of the PLCs by stipulating that the secretary of a PLC must be a deputy party chief or a member of the standing committee of a party organization at the same level.[28]

Starting in the mid-1990s, the party gradually increased the personnel and funding of subnational PLCs and expanded their responsibilities, empowering them to better formulate, coordinate, and implement social control and political repression.[29] In 1995 the General Office of the CCP Central Committee directed the PLCs to "organize and coordinate the work of social comprehensive management of law and order and public safety."[30] And in 1999, the CCP formed the 610 Office within the PLCs at all levels, further solidifying their status as the party's domestic security taskmaster.

A key measurement of post-Tiananmen reliance on the CPLC is the regularity of its national conference on political-legal work. In the 1980s, the party convened only three such conferences; since 1992, the conference has been held on an annual basis. Unlike in the 1980s, when neither Deng nor General Secretary Hu Yaobang attended the gatherings, top leaders such as Jiang Zemin and Xi Jinping have attended and addressed the conferences.

In the 2000s, the power of the PLCs in domestic security affairs was augmented further. In November 2003, as part of its effort to strengthen its leadership of public security agencies, the party mandated that directors of local public security agencies be members of the standing committee of the local party committee or else a deputy governor or mayor. Since the heads of the local PLCs are also members of the local CCP standing committees, it became a standard practice to appoint the same person to head both the local PLCs and the PSBs. With one stroke, the party elevated the political status of the local PSBs and gave the local PLCs an unprecedented direct role in public security.[31]

This coincided with the elevation of the head of the CPLC to the Politburo Standing Committee and with the tenures of hardliners Luo Gan and Zhou Yongkang—a capable and ruthless apparatchik—atop the CPLC. Under Luo and Zhou, the CPLC and the local PLCs acquired more responsibility to maintain "social stability." The landmark November 2003 CCP document on domestic security pledged to increase funding, and that is precisely what happened: the domestic security budget, which funds the agencies under the supervision of the CPLC and the local PLCs, grew rapidly.

After Xi became party chief in November 2012, the political status of the CPLC apparently declined. In late 2013 Zhou Yongkang was placed under investigation for corruption and was eventually convicted for having taken millions of dollars in bribes, as well as for leaking state secrets. He was sentenced to life in prison. His successor, Meng Jianzhu, was a Politburo member but not a member of its standing committee. Indeed, the role of the CPLC and the PLCs was reduced even before Zhou was purged. At the local level, the practice of appointing the heads of the PLCs to lead the PSBs was gradually phased out after 2012.[32] In November 2013, the party announced the establishment of the Central National Security Commission, personally chaired by Xi.[33] Given its domestically focused mandate, this new organ, at least in theory, could assume many of the domestic security responsibilities currently assigned to the CPLC.

In practice that has not happened, although Xi's commission perhaps still functions as a sword of Damocles over the CPLC. For now, despite the presence of the National Security Commission, the CPLC continues to be the party's principal overseer and coordinator of domestic security. Frankly, to replace the existing bureaucracy would be an immense organizational challenge; notably, the new local national security offices are all housed inside local PLCs. Although it is possible that eventually the party may expand the power of the local national security offices and turn them into bureaucracies independent of the local PLCs, it is equally possible that future local national security offices will be no more than relabeled PLCs.

It does not appear that Xi's goal is to replace the CPLC and its local affiliates, so much as to assert more direct control over them—the constant concern of a dictator seeking to maintain the political loyalty of his repressive apparatus. Thus in the late 2010s, years after creating his own national security commission, Xi was still focused on the future of the PLCs, purging the central and local bureaucracies and, in January 2018, announcing the CPLC's reorganization. The campaign has claimed many senior public security officials as well as rank-and-file police.[34] According to Chen Yixin, Xi's point man in charge of the purge, by June 2021, 12,576 police officers had turned themselves in and nearly 100,000 officers were investigated and punished for "violations of discipline and law."[35] This may be the most thorough purge of the coercive apparatus since the end of the Cultural Revolution. In addition, in January 2019 Xi issued an important document, "CCP Rules on Political-Legal Work," formalizing existing practices and cementing the party's supremacy over the political-legal sector. A key objective of this document is to extend the PLCs even deeper into rural China. Whereas historically counties were the lowest subnational level at which PLCs operated, now they have been established at the township level.[36]

In other words, Xi has not weakened the political-legal sector. He has domesticated it, while apparently hoping to press it that much further into the lives of citizens by installing PLCs in China's tens of thousands of townships.

Organization and Functions of the CPLC and the PLCs

Despite the vast domestic security apparatus under their supervision, the CPLC and the local PLCs are not themselves large organizations. The current staff size of the CPLC is not publicly known, but because the CPLC's authorized headcount in the mid-1990s was only fifty, it is fair to speculate that its entire staff today is in the low hundreds. As for the local PLCs, we can gain a sense of their scale by drawing from what is known about specific committees. In 1983, Ji'nan's PLC had only three sections: the secretariat, research section, and political work section, which oversaw official appointments in the political-legal sector. By November 1996, the Ji'nan PLC had been reorganized and had five sections: a general office, a research and propaganda section, a section responsible for supervision of law enforcement, a section devoted to comprehensive management of public order, and a cadres section, which oversaw personnel decisions. Expansion and added responsibilities almost certainly demanded increased staff. Additional sections, such as the 610 Office and stability-maintenance office, were added in the early 2000s. In this case, we know that manpower increased: Ji'nan's PLC grew from thirty-two officials and staff in 2002 to forty-six in 2009.[37] Likewise the PLC of Yuanmo County, in Yunnan Province, grew from just five people in 1995 to eighteen in 2010.[38]

The sizes and organizational charts of PLCs vary across jurisdictions. In 2020, the PLC of Wuhan, a large city, had sixty-nine officials and staff.[39] By comparison, most county-level PLCs in the late 2010s had about ten to twenty officials and staff.[40] A locality's population and financial resources likely determine the size of its PLC. The PLC of the 140,000-population Yuwangtai District in the city of Kaifeng had a staff of only six in 2018, while its counterpart in Tianchang, a city in Anhui Province with roughly four times the population of Yuwangtai, had twelve full-time officials and staff persons in 2019.[41] (For the most part, districts are at the same level of the administrative hierarchy as counties, but whereas counties typically collect multiple rural townships and villages under their umbrella, districts are usually carved from large cities.) Wealthy and politically important jurisdictions have relatively large PLCs. Miyun, a district of Beijing with a population of half a million, had an authorized establishment of fifty-one staff and additional support staff of seventeen in 2018.[42]

No information about the internal organization of the CPLC can be found on its website, although we can make informed guesses as to how its work may be divided.[43] According to the March 2018 CPLC reorganization plan, the commission was to share with the MPS responsibility for repressing "evil cults," thus warranting a section dedicated to that cause. Another responsibility assigned to the CPLC was to coordinate, push, and supervise comprehensive law and public-order management, a task probably requiring its own stand-alone department. The reorganization abolished the Central Stability Maintenance Leading Group and its office and transferred this responsibility to the CPLC, which most likely entailed the establishment of a department responsible for social stability. It should be noted that the transfer of these responsibilities to the CPLC did not actually expand the power of the CPLC, because the offices and commissions abolished by the reorganization were located inside the CPLC and staffed by its personnel. Thus the "transfer" merely streamlined the CPLC's management structure and eliminated unnecessary shell departments.

Since the subnational PLCs often mirror the organizational arrangements of the CPLC, we may also gain some insights into the internal organization of the CPLC by looking at its lower-level counterparts. Among these, the PLC of Guizhou is one of the few provincial-level PLCs that reveals details of its internal organization. According to its website, the Guizhou PLC has fifteen sections and offices. They include thirteen divisions: cadres; propaganda, including cyber administration; two divisions for supervising implementation of the law; four for comprehensive management; two for stability maintenance; one for supervision and investigation of information, which likely means vetting law enforcement tips; one for crime prevention; and one responsible for reforming "cultists." This PLC also has two offices: a general office and a research office.[44] The responsibilities of these many divisions are easier to understand in terms of the five substantive areas they work on: management and vetting of officials in the political-legal sector (cadres), conventional law enforcement and public safety (comprehensive management and information investigation), regime security (stability maintenance and intelligence), coordination of political-legal work (supervising implementation of the law), and suppression of banned religious groups. It is likely the CPLC's divisions also fit into these five groupings.

An examination of the websites of local PLCs and yearbooks published by local governments shows that cities and counties have wide discretion in deciding the internal organization of their PLCs.[45] A typical local PLC in the late 2010s would have been led by a party secretary, assisted by two deputy secretaries. Together, they oversaw six or seven functional offices, which might have included a general office handling administrative affairs, an office for comprehensive management of law and order that dealt with routine public safety and law enforcement issues, an office responsible for supervising the legal system, an office in charge of vetting and evaluating officials in the legal system, a stability maintenance office (usually the office with the largest staff), a 610 Office, and a propaganda office. Some PLCs had additional offices responsible for national security, anti-terrorism, anti-narcotics, and "anti-infiltration" activities. Judged by the names of these offices, it seems that those in charge of social stability, anti-"cult" operations, anti-infiltration operations, and national security directly supervised and coordinated the work of the surveillance state.[46]

Much as the head of the CPLC is a high-ranking politician—formerly a member of the PSC, now still a member of the Politburo—so too is the head of a local PLC. Specifically, the secretary is a member of the standing committee of the local party organization, akin to the PSC but with purely local authority. Other members of the local PLCs include the police chief at the relevant level, the president of the local court, the head of the local procuratorate, and the director of the local justice bureau (responsible for prisons and the legal profession). If a county or district has an MSS office, its director will also be a member of the relevant PLC.[47]

Activities of the Local PLCs

Given the small size of local PLCs, it should be unsurprising that they are not directly involved in law enforcement, public safety, surveillance, and social stability. Their job is rather to supervise and coordinate the activities of those directly responsible—courts, the procuratorate, and PSBs. PLCs exercise the authority of local party chiefs, which means that they also are driven by political incentives. By giving the PLCs power to screen the appointment and promotion of law enforcement officials, the party has created a potent instrument for ensuring compliance and cooperation of officials in the coercive apparatus. Even more important is the alignment of the political incentives of local party chiefs with the mission of the PLCs. As maintaining stability became a top political priority for the party in the post-Tiananmen era, local party chiefs could ill afford to be negligent in matters of domestic security.

An important responsibility—and opportunity—for local PLC officials lies in the work conferences. The two-day CPLC conference is attended not only by the principal central government officials responsible for security and law enforcement—as well as relevant figures from state-owned enterprises—but also the heads of the provincial PLCs, who thereby demonstrate their importance to the party. As we have seen, this conference is followed by additional, similar conferences at the provincial level. These are attended by the heads of the municipal and prefectural PLCs, which hold their own annual work conferences shortly thereafter. Based on this schedule, by the end of February of each year, the party's annual domestic security agenda has been fully communicated to the county and district levels.

An examination of the reports of some local PLCs shows that their responsibilities fall into five categories:

(1) Domestic security tasks, including crackdowns on banned spiritual organizations, coordinating elevated security measures during sensitive periods, and identifying and surveilling key individuals. These include veterans of the People's Liberation Army (PLA), who tend to be well organized in their demands for better pensions and other benefits; members of banned spiritual organizations; and petitioners—ordinary citizens who, dissatisfied with their treatment by officials, bring their grievances to higher authorities. Additionally, a key security task involves dealing with mass incidents: protests, riots, strikes, and large-scale petitions, which can see throngs of people gathering at public offices.

(2) Screening officials working, or seeking work, in the political-legal sector for political loyalty.

(3) Overseeing the buildout of the domestic security infrastructure.

(4) Coordinating resolution and disposal of major legal cases.

(5) Overseeing law enforcement, public safety, and anticrime and security campaigns.[48]

While local PLCs supervise the courts and procuratorates, intervene in major legal disputes, and oversee periodic anticrime initiatives, their primary task is to ensure that local manifestations of the coercive apparatus and other state bureaucracies implement the party's domestic security agenda. The ability of the PLCs to fulfill this mission rests on their capacity to organize and coordinate campaigns, take security measures during sensitive periods, process information and intelligence, and facilitate the flow of information laterally and upward. It is a marker of the PLCs' reliability that the party has depended on them heavily to build the main components of the techno-surveillance state, such as grid management and the Sharp Eyes program.

PLC activity reports suggest that their direct involvement in operations of the coercive apparatus most commonly includes supervision and coordination of extra security measures during sensitive periods.[49] This particular task receives a high priority because local officials are likely to be punished severely if a major incident occurs on a politically charged anniversary or during an important event. Thus the municipal PLC of Tianjin, for example, oversaw enhanced security measures during the annual "Two Sessions"—the gatherings of the National People's Congress and the Chinese People's Political Consultative Conference, in this case in 2017. The committee convened a video conference every evening and received daily reports from the police and other agencies to coordinate security efforts. Its senior officials also personally visited key government offices, checked on their security measures, and inquired about the surveillance of key groups. In September, shortly before the National People's Congress, the Tianjin PLC held a meeting to mobilize the city's security forces, cyber censors, and other local departments. Following the meeting, the committee supervised implementation of extra security measures, such as initiatives designed to prevent PLA veterans from staging protests and to protect critical infrastructure and Congress venues.[50] Local PLCs in other jurisdictions report engaging in similar activities during sensitive periods.[51]

Local PLCs trumpet their coordinating role in their annual reports. The PLC of Wuhu, a city in Anhui province, claims in its 2010 annual report that it coordinated the public security work and promoted the resource sharing across government agencies. A late 1990s Shenzhen PLC report boasts of "coordinat[ing] the work and covert political security operations of the local state security and public security agencies." Further, the PLC claims to have "collected a large quantity of important intelligence and information, successfully carried out effective surveillance of hostile elements who intruded into the city, . . . delivered a blow to foreign religious groups trying to infiltrate China," and, likely referring to underground labor unions, "thwarted a conspiracy to organize 'Solidarity-style' illegal labor groups."[52]

We also know of local PLCs supervising the surveillance of key targets. In 2000, the district PLC of Longgang, in Shenzhen, oversaw a large investigation of the district's Falun Gong followers, which was conducted by local public security and state security agencies. The PLC claims to have formed three-person teams to carry out close surveillance of key Falun Gong members, some of whom were detained during sensitive periods. In Shanghai's Chongming County in the early 2010s, the PLC engaged in similar supervisory activities involving many government agencies.

In general, recruiting informants and other auxiliary security personnel is a routine task that local PLCs oversee.[53] PLCs also facilitate implementation of China's Safe City project, a loosely defined public safety initiative combining investments in technology, manpower, infrastructure, and propaganda to reduce crime, traffic accidents, incidents of fire, and other hazards. The PLCs coordinate funding, run the propaganda campaign, and evaluate and certify Safe City projects.[54]

For a granular look into the operations of a local PLC, I turn to a log recording the late December 2015 work of the PLC of Futian District in Shenzhen.[55] There was nothing special about this week; the log just happens to be available, and it provides an up-close encounter with a week in the life of a typical PLC. During the week of December 22–29, the stability-maintenance office of the district PLC worked with the municipal letters-and-visits bureau to resolve the issue of petitioners who alleged that they had been defrauded by an online fundraising platform. The same office coordinated the handling of some of Shenzhen's PLA veterans who were planning an activity that had the potential to trigger unrest. The office also kept an eye on disputes between labor and management at two local companies. The PLC compiled reports for the Shenzhen Municipal Stability Maintenance Office about its daily stability-maintenance operations and its accomplishments in cracking down on protestors and petitioners. Meanwhile, the Comprehensive Social Management Office of the PLC inspected the implementation of antiterrorism security measures, ordered a crackdown on trafficking in wild animals, and prepared for a major propaganda event on traffic safety. The PLC's 610 Office assisted the head of the Shenzhen State Security Bureau during the latter's research visit to the district, supervised two investigations of "cult" activities, and issued two documents certifying that particular citizens were not associated with "evil cults"—documents they were required to obtain before securing employment.

Thus local PLCs perform many different tasks, ranging from intervention in mundane civil disputes to preemptive measures against potential subversives. For a regime seized by paranoia and insecurity, a versatile PLC is worth every yuan.

COMPARED WITH OTHER DICTATORSHIPS—even communist regimes in the former Soviet bloc—the CCP stands out for its achievements in building a sophisticated infrastructure of command, control, and coordination. For the most part, this infrastructure came about after 1989, as top leadership prioritized regime security and as the rapidly growing economy produced resources enabling the new system.

Even in the wake of Xi's purges and reforms, the organizational linchpin of the Chinese surveillance state is, without a doubt, the political-legal committees, with the CPLC at the top. Other domestic security entities may have more senior party leaders and boast more exalted titles, but it is the PLCs, a specialized party bureaucracy, that perform the two roles essential in addressing the coercive dilemma and the challenge of coordinating surveillance across security agencies and assorted state and nonstate actors. First, the PLCs vet key personnel in the coercive apparatus, enabling the party to appoint and promote loyalists to important positions in the police and secret police. Second, the PLCs coordinate the implementation of the party's security agenda on a routine basis, making it possible for China's system of distributed surveillance to function effectively.

The increased political heft and security responsibilities of the party's PLCs at all levels illustrates the CCP's skillful application of Leninist organizational principles in confronting emerging threats to its power. To be sure, the party's system of command, control, and coordination is by no means perfect, and frequent revelations of fraud and venality in China's coercive apparatus, including among senior officials in the PLCs, show that China's rulers have not solved the problem of corruption—another challenge that plagues all dictatorships. Yet it is difficult to imagine that the party could have maintained its grip on power as successfully and ruthlessly as it has in the post-1989 era without the PLCs as domestic security taskmasters.

CHAPTER 3

Organizing Surveillance

China's vast coercive apparatus includes the People's Liberation Army; the People's Armed Police; the Ministry of Public Security (MPS), which oversees a police force of about 2 million officers (as of 2010); the Ministry of State Security (MSS); the People's Militia; and an auxiliary police force probably larger than the uniformed police.[1] However, only a small number of institutions in this apparatus play a direct role in the operations of the surveillance state. As far as I can determine, three institutions form the operational core of the surveillance state: the Domestic Security Protection (DSP) unit of the police, regular community police stations, and the MSS and its local outfits. Taking direction from the political-legal sector, which itself receives marching orders from the highest ranks of the CCP, these three institutions carry out China's distinctive approach to coercion.

Many features of the coercive apparatus itself are notable. For one thing, it is small relative to the size of the Chinese population; the country is underpoliced by international standards.[2] Even more strikingly, China does not have the equivalent of a Stasi or a

KGB—a powerful secret police agency that conducts espionage operations abroad and counterintelligence and political spying at home.[3] In China, these three functions—foreign espionage, counterintelligence, and surveillance of subversive elements—are split between two different security bureaucracies. Before the establishment of the MSS in 1983, the division of labor for the collection of external intelligence, counterintelligence, and domestic surveillance actually resembled that in the democracies more than that in the communist dictatorships. The Central Investigation Department, a party bureaucracy, was in charge of external espionage—similar to the US Central Intelligence Agency or Britain's MI6. Meanwhile the MPS was tasked with counterintelligence and domestic surveillance, functions performed by the Federal Bureau of Investigation in the United States and MI5 in Britain.

The MSS is more similar to a Stasi or KGB but still is not quite the same. It spies abroad and handles counterintelligence at home, but it has limited purview with respect to domestic surveillance, focusing on ethnic minorities, foreigners, and individuals with overseas connections. It is the MPS, the domestic security agency responsible for the frontline police, that maintains the dominant role in domestic surveillance. Furthermore, the MPS is more powerful than the secret police, as indicated by the political status of the minister of public security, which exceeds that of the minister of state security.[4] Within the MPS, domestic surveillance responsibilities are further divided between the Domestic Security Protection unit, which takes the lead on matters of political security, and the much larger regular police force, with assistance from other state entities, CCP members, and civilian activists.

Several factors may explain the absence of a "full-service" secret police in China. One is elite rivalry over bureaucratic fiefdoms.

Historically, Zhou Enlai, the long-serving premier, sought to maintain personal control over the party's intelligence arm (the Central Social Affairs Department and its successor, the Central Investigation Department) during both the revolutionary years and the pre–Cultural Revolution period of the People's Republic.[5] This interest could reasonably have precluded his agreeing to the formation of a full-service secret police. The concern was that Zhou might lose control of the Central Investigation Department had it also assumed responsibility for domestic spying. Such an empowered secret police would have aroused Mao's suspicions, encouraging him to appoint a loyalist to oversee the combined secret police. Zhou was not that loyalist; fearing estrangement from his role as intelligence chief, he had ample reason to preserve the status quo rather than seek more responsibility for his agency.

Zhou's interests may therefore have overlapped with Mao's. As a dictator known for paranoia and disdain for bureaucracy, Mao might have felt that a powerful secret police agency could threaten his power. Hence, it was better to divide security responsibilities among rival bureaucracies. Additionally, Mao had ideological reasons to oppose an empowered secret police. A champion of the so-called mass line—mobilizing the people to carry out the regime's tasks—Mao distrusted elite institutions. In 1951 instructions to the MPS, Mao decried the practice of "secrecy" and endorsed "mass mobilization under the party's leadership" to implement the campaign against counterrevolutionaries.[6]

The difference between organization of the surveillance states in China and the Soviet Union is all the more striking because China opted for its own model in spite of Soviet influence. Indeed, Moscow dispatched security experts to China as soon as the People's Republic was founded in 1949 and continued sending agents thereafter. Yet the Soviet agents had little impact in shaping the basic structure of China's domestic security apparatus.[7] China may have imported Leninist institutions from the Soviet Union, but the Maoist regime was far more invested in mass mobilization as an instrument of rule, and in the repression and surveillance of political threats in particular. Practically speaking, reliance on the masses enabled the Maoist regime to keep the size of the formal coercive apparatus relatively small.

As illustrated below, decisions taken during the early days of the Maoist regime shaped the future architecture of the formal surveillance state. In particular, the decisions prefigured the development of a multilayered organizational structure and multiple security agencies with distinct remits—all of which collectively guard the party's political monopoly. Compared with dictatorships that rely on a powerful full-service secret police agency, the Chinese system seems to have several advantages: it costs less, is more effective in stymying potential opposition, and poses fewer dangers to the top leadership of the regime.

The Domestic Security Protection Unit

The Domestic Security Protection unit (*guonei anquan baowei*, often abbreviated as *guobao*) shoulders the primary operational responsibility of the Chinese surveillance state. Brief references to its activities are readily available in local yearbooks, but there has been no scholarly study of the DSP. Housed within the MPS at the national level, and the Public Security Bureaus (PSBs) at the local level, the DSP was known as the political security protection unit (*zhengzhi anquan baowei*) prior to 2000. In 2019, the Chinese government restored that name, most likely to emphasize the unit's primary responsibility for guarding against threats to CCP rule.[8] However, to avoid confusion, I use the terminology of Domestic Security Protection/DSP throughout this study.

Within the MPS, DSP is known as the ministry's first bureau, a designation indicative of its status. The first bureau is considered the most important and most powerful department in the MPS, and, according to a former MPS officer, its director invariably is promoted to the position of vice minister of the MPS.[9] There are also DSP units within provincial, municipal, and county or district PSBs.

Official sources provide few details about the sizes of local DSP units; as usual, one must glean from disparate data sources in order to assemble a broader picture. In the early 1950s, about 40,000 police officers were tasked to the MPS political protection units, the predecessor to DSP. These comprised about 10 percent of the total police force.[10] Today, DSP units are relatively small and probably account for around 3 percent of the police force. In 1985, Xinjiang had only 618 police officers assigned to DSP units across the region's 106 jurisdictions.[11] In Puyang, a midsized city in Henan Province, the municipal DSP unit had forty officers in 1993, just 2.4 percent of the force.[12] Other jurisdictions also report small DSP units. The DSP unit in Shuimogou District, in Urumqi, had twenty-four officers (5.7 percent of the total police force) in 2012. By comparison, the criminal investigation unit of the district PSB had forty-eight officers, or 11.4 percent of the force. In Zhuzhou, a midsized Hunan city, the municipal DSP unit had only twenty officers during 1997–2000; the DSP unit in Daqing's Sartu District had just six officers in 2002–2004; the same unit in Zaoyang County had eight officers in 2012; and Wuhan's municipal DSP unit had 216 officers in 2014, assigned to a population of more than

10 million.[13] If we assume that the total number of DSP officers nationwide is about 3–5 percent of the police force, as the limited local data suggest, China's domestic secret police number around 60,000–100,000, or one DSP agent for every 14,000–23,000 people. Recall that, in 1989, the East German Stasi had one full-time employee for every 165 people, while the former Soviet Union had one KGB officer for every 595 people.[14] Even if only half of Stasi and KGB employees were responsible for domestic spying, the share of domestic security agents per population was at least forty times greater in East Germany than in China and eleven times greater in the USSR.

It is notable that the CCP did not decide to dramatically increase the size of the domestic secret police when resources allowed. During the period of Mao and Zhou Enlai, the state could not have assembled the Chinese equivalent of a Stasi even if it had wished to. Yet with the increased post-1989 investments, the size of the DSP nonetheless remained small. In practical terms, CCP leaders probably understand that they do not need a large domestic secret agency because less elite frontline police can perform most routine surveillance tasks. And, again, the party has an interest in preventing a spying agency from becoming a locus of power that can threaten its own political supremacy.

We may infer from the small size of the DSP units in counties and districts that only relatively large provincial and municipal DSP units possess substantial operational capabilities—such as the capacity to conduct high-priority investigations, surveil key targets, and engage in sophisticated intelligence gathering and analysis. The primary function of county and district DSP units appears confined to recruiting informants and directing community-based police officers to perform routine surveillance tasks. Their secondary function is to execute orders from superior DSP units and local party organizations. Local DSP units thus serve two masters. They are part of a functional hierarchy and therefore are supervised by superior security agencies. But at the same time, they also rely on local governments for funding and personnel, and they must respond to the needs of those local governments.[15] An effect of this dual role is that district and county DSP units tend to perform a small number of high-priority operations and delegate routine tasks to local police.

Roughly once every three years, the MPS convenes national conferences on domestic security to issue directives, provide operational guidance, and facilitate the exchange of ideas among subnational DSP units.[16] The government (probably through the MPS) also holds specialized conferences on "political investigation work," most likely to develop and share with local DSP units techniques used in surveillance and investigation of individuals and organizations deemed political threats.[17]

The Wenbao Unit

Adjacent to the local DSP units responsible for surveillance of the general population are cultural and economic protection (*wenbao*) units, which handle surveillance in educational and academic institutions. At the national level, the MPS does not have a separate wenbao department. Instead, it has a university division (*daxuechu*). Some provincial PSBs, such as that in Zhejiang, follow the MPS model and have a university section inside their DSP division. In other locations, such as Beijing, there is a stand-alone wenbao unit. PSBs in some prefectural-level cities have stand-alone wenbao *zhidui* (regiments).[18] Other prefectural-level PSBs—such as those of Bingzhou, in Shandong, and Zhengzhou, in Henan—have a wenbao *dadui* (battalion) within their DSP unit.[19]

District and county-level PSBs may also have wenbao units. As most universities and major academic institutions are located in prefectural-level cities, responsibility for maintaining security and surveillance in these institutions belongs to the wenbao units in the municipal PSBs.

Wenbao officers use the same surveillance tactics as their DSP colleagues. As part of their routine tasks, wenbao officers visit university campuses to advise on and inspect stability-maintenance work, such as surveillance of ethnic-minority students.[20] Yearbooks for the city of Xi'an, which has many major universities, report that the wenbao unit of the municipal PSB actively collects information and intelligence, operates a network of informants, provides intelligence to the police, surveils and controls "key individuals," investigates "enemies," suppresses illegal religious activities and "infiltration" by foreign and domestic NGOs, and handles mass incidents in the city's cultural and educational institutions. The wenbao unit convenes a quarterly meeting covering security on university campuses, which representatives of the city's institutions of higher education presumably are required to attend.[21] The yearbooks of the Wuhan PSB reveal similar operations.[22]

Prominent dissidents have confirmed that wenbao officers were responsible for their cases and would meet with them regularly, often over tea and meals. A professor in Shanghai reports that, as far as he is aware, wenbao officers exclusively handled his case. Xu Youyu, a political philosopher and former research center director at the Chinese Academy of Social Sciences (CASS), has said that both the DSP and the municipal wenbao managed his case, but the wenbao unit had primary responsibility. Teng Biao, a human rights activist and lawyer who was a faculty member of the China University of Political Science and Law, was surveilled by the wenbao unit of the Beijing PSB, with the cooperation of the DSP unit of Changping District, where he resided.[23]

Investigative and Intelligence Operations

According to an authoritative textbook on policing, the role of the DSP unit is to collect and analyze intelligence, detect and act against individuals who endanger social and political stability and national security, target religious and ethnic groups, strengthen security in academic institutions and state-affiliated entities, and recruit informants.[24] The mission statement of the DSP battalion of Shicheng County, in Jiangxi Province, likely reflects the tasks typical of a local DSP unit: its announced role is to "collect, grasp, process, and conduct research on intelligence and information related to social and political stability and national security; provide opinions and measures; and organize, investigate, control, safeguard against, and handle cases and incidents that harm social and political stability and national security and unity." With these goals in mind, the DSP unit also surveils banned religious groups and performs antiterrorism duties. Not directly mentioned is that the DSP unit recruits informants from all social strata.[25]

Based on summaries of the activities of the DSP units in local yearbooks, their most common and most important task is to investigate individuals and organizations that pose potential threats to the CCP, as well as incidents involving these potential threats. Although an average county-level DSP unit has insufficient manpower to carry out routine surveillance, it can undertake special investigations for which the regular police are ill-equipped due to a lack of investigative skills and competing demands on their time. Targets of special investigations include leaders and activists who organize and participate in uprisings, riots, and protests; hostile foreigners engaging in infiltration and sabotage; illegal organizations and publications; organizations and individuals engaging in ethnic-separatist activities; leaders and activists in secret societies; individuals who collude with external forces; and organizations and individuals engaging in terrorism.[26]

Disclosures of DSP activities by local PSBs confirm that these units pursue most of the above targets. The public security gazette of Zhuzhou, in Hunan Province, provides a relatively detailed description of the investigations and operational accomplishments credited to its DSP unit in the 1990s. During the decade, the municipal DSP unit conducted investigations on university campuses that uncovered individuals suspected of participating in illegal organizations. Its investigations also targeted foreign teachers and members of NGOs. In 1994, with the aid of "technical means" and "inside informants," the unit tracked the activities of a student leader from the Tiananmen movement and put pressure on his commercial activities to prevent him "using business to fund politics." In 1997, the unit shifted its focus to religious groups, and in 2000 it carried out secret investigations of various qigong organizations.[27]

Reflecting the CCP's constant fear of organized opposition, DSP units prioritize illegal and unregistered religious and political groups, which may operate in secret. In the Sartu District of the city of Daqing, in Heilongjiang Province, the DSP unit spied on religious practitioners and members of secret societies, even turning some of the practitioners into informants in the late 1980s.[28] In the 1990s and early 2000s, the DSP unit of Shulan County, Jilin Province, spied on secret societies and religious groups and arrested their leaders.[29] Similar investigations of religious groups can readily be found in summaries of the accomplishments of the DSP units in most jurisdictions since the early 1990s, which boast of "smashing" so-called cults.[30] Foreign NGOs are also priority targets. In 2016, the DSP unit of Hanyuan County, Sichuan Province, investigated several foreign NGOs. Wuhan's DSP unit claims to have conducted operations in 2009 and 2013 to detect infiltration by external groups.[31]

These reports similarly indicate that DSP units investigate a range of individuals deemed political threats. The DSP unit of Xishuangbanna Dai, an autonomous prefecture bordering Laos and Myanmar and home to many ethnic minorities, reports that in 2010 it investigated more than a thousand Uighurs entering the prefecture and uncovered groups smuggling Uighurs out of China. In Miyi County, Sichuan Province, in 2008, the DSP unit reported having conducted special investigations of dismissed private school teachers, army veterans, and migrant workers upset about unpaid wages. It was thought that these and other key individuals might cause trouble during the Two Sessions and the Beijing Olympics.[32]

The DSP collects intelligence not only in order to stymie protests and pursue arrests but also to build a database of known and potential political threats. It is likely that the units currently follow instructions provided in the MPS's 2002 "Opinion on Conducting Basic Investigations of Those Who Are Subjects of Domestic Security Protection Work." This document is not publicly available, but there is reason to believe it establishes protocols and priorities for DSP collection of essential personal data, including information about physical features of individuals, family background, residence, employment, and social connections.[33] Disclosures in local yearbooks provide examples of how such information has been gathered over the years, including before the 2002 instructions. The DSP unit of Zhuzhou, Hunan, conducted such "basic investigations" as early as 1991, when the provincial PSD held a conference on "covert struggle" and issued a document on identifying political threats. The municipal DSP unit implemented what it called the '91 Project, a five-month investigation resulting in the designation of more than a thousand individuals as targets. Their information was subsequently entered into a police registry. In addition, the project designated about two dozen target organizations; these were entered into a "registry of organizations."[34]

Alongside dragnets, DSP units conduct more focused investigations. For example, in 2005, Wuhan's DSP unit investigated foreign NGOs in the city.[35] Beijing's DSP unit spent 2000 exclusively targeting illegal organizations.[36] A 2006 reference by the Beijing DSP unit to specialized databases of "evil cults" and "antiterrorism information-management systems" speaks to the targets of their surveillance activities and the sorts of intelligence they collect and process.[37]

Some of this intelligence is gathered by informants who monitor targeted individuals and groups and seek out public reactions to state policies and major social trends. The intelligence that informants provide DSP units is divided into three categories: enemy intelligence (*diqing*), political intelligence (*zhengqing*), and social intelligence (*sheqing*). (In some localities, political and social intelligence are combined into one category.)

According to two textbooks on public security, enemy intelligence consists of information about hostile groups and their activities. Especially valued is intelligence about those groups that "seek to gain intelligence, instigate defections, engage in sabotage, endanger state security, subvert state power, undermine national unity, and fuel armed revolts and harassment." Enemy intelligence also includes information about collusion and secret contacts among hostile foreign and domestic forces, infiltration by hostile forces, and activities endangering social stability. Information about major criminal activities, such as drug trafficking and organized crime, is also classified as enemy intelligence.

Political intelligence refers to "information about political trends and developments within and outside Chinese borders that influence or may influence domestic social and political stability and national security." This may include "reactions by people of various strata to party and government policies, laws, and major domestic and foreign events." Although political and social intelligence is difficult to distinguish under some circumstances, the two textbooks define social intelligence as information "about various destabilizing factors that exist within society or that are likely to influence social and political domestic stability." Also included in the social category is "public opinion concerning major accidents, natural disasters and strikes, especially reactions by 'representative individuals,' and notable social trends."[38]

Despite secrecy surrounding recruitment, publicly available sources frequently refer to the alleged successes of DSP informants.[39] Indeed, information about both the number and the output of DSP spies and informants is readily available in local yearbooks. (As discussed in more detail in the later chapters, spies are recruited by police, and their identities are protected. Informants are recruited by political authorities, and their identities may be protected or may be known to the public.) In general, a DSP unit in an average city or prefecture collects several hundred pieces of information and intelligence that are deemed relevant to regime security each year.[40] In 1998, the DSP unit in Zhuzhou, Hunan Province, claimed to have made "friends" and established "working relationships" with 249 individuals, in addition to recruiting 65 people explicitly described as spies.[41] Between 1998 and 2002, the

DSP unit in Shulan, Jilin Province, signed up an average of 163 informants per year, in addition to an unspecified number of spies.

As for the findings of informants and spies, DSP units routinely discuss specific information collected by their agents and used to intervene against dissidents. The Miyi DSP acknowledges that, in 2008, its informants helped police gain "timely information about Miyi Middle School teachers who were attempting to organize a strike via the internet." Informants also assisted the police in obtaining "extensive information about a plan to present a collective petition and an attempt to block construction at the site of a new commercial development."[42]

Routine Operations and Direct Repression

DSP units frequently deputize frontline police in the implementation of routine surveillance. This arrangement not only frees up the limited manpower of DSP but also effectively expands the reach of the secret police without increasing its size. In 1991, in the Jilin Province city of Panshi, the municipal DSP unit designated officers responsible for political policing in each of the city's police stations (*paichushuo*). All frontline police in the city were required to undergo training in political security, which presumably was provided by the DSP unit.[43] In 2000, the Beijing PSB promulgated "Protocols for Domestic Security Protection Work by Police Stations"; the document itself is secret, but it is known to specify tasks and evaluation criteria for police stations undertaking DSP work.[44] The protocols require that police stations establish information-collection networks in communities, improve monitoring of key individuals, and maintain "control" over such individuals. In 2016, the DSP unit of Yueyanglou District, in Yueyang, Hunan Province, worked with frontline police to step up surveillance of "key targets."

Police were instructed to keep the DSP unit aware of targets' movements and activities. In 2012, frontline police in Leshan, Sichuan, provided the city's DSP unit daily updates on "factors of instability."[45]

Alongside frontline police, DSP units work with other government agencies to coordinate surveillance and security operations. In 2005, Beijing's DSP unit reported collaborating with the city's customs office, education commission, and tourism bureau to set up a regular coordination mechanism that would "guard against infiltration by foreign religious groups." In Henan, the DSP unit worked with the Neihuang County United Front and with the Religious Affairs Bureau to carry out "safety inspections" of religious sites during major holidays and other periods.[46]

Despite their relatively small size, local DSP units conduct direct surveillance of priority targets and participate in security operations, such as breaking up illegal religious gatherings, confiscating illegal religious materials, arresting political suspects, and suppressing protests. In 2012, the DSP unit in Weng'an County, Guizhou Province, surveilled key individuals and army veterans.[47] Between 1999 and 2003, the Panshi DSP unit carried out annual operations against illegal religious groups, including Falun Gong, surveilling their venues and practitioners. The DSP unit in Taijiang District, Fuzhou, highlights its surveillance and investigations of illegal religious groups between 1991 and 2005, operations that included breaking up gatherings, confiscating materials, and arresting members.[48]

The operational role of local DSP units in preventing and quelling protests reflects post-Tiananmen priorities. Indeed, the DSP units have been more involved in controlling protests since the late 1990s, when stability maintenance became a high priority.

These operations usually involve the deployment of informants, who infiltrate suspected protest groups and report to police about their plans. Wuhan's DSP unit reports that, in 2013, it "effectively used intelligence to gain advance warnings about mass incidents" and "successfully handled more than a hundred" such incidents.[49] DSP units usually use brief, euphemistic language like "handle" (*chuli*) in discussing these episodes. Regardless, the incidents speak to the important role of DSP units in helping local authorities suppress social unrest.[50]

Police Stations

The frontline of China's law-enforcement system is the police station, sometimes referred to in scholarly literature by the acronym PCS—a reference to the Chinese term, *paichushuo*. Police stations perform most routine law enforcement and public security functions, the sphere of which has grown since the mid-1950s. According to the "Rules on Organization of Public Security Police Stations" issued in 1954 and technically in force until 2009, police stations were supposed to enforce laws and rules of public order, suppress sabotage activities by counterrevolutionaries, prevent crimes and halt crimes in progress, maintain control over counterrevolutionaries and other criminals, enforce the hukou, and provide guidance to local security committees.[51] During the post-Mao era, and in particular in the late 1990s, frontline police gradually took on more law enforcement responsibilities and assumed more active roles in the surveillance state. The pivotal event was the MPS's 1997 national conference on "Domestic Security Work at the Basic Level." There the MPS announced that "maintaining social stability" would now be a key police mission; police stations were required to set up domestic security offices or appoint officers dedicated to domestic security duties.[52]

The assignment of surveillance duties to the frontline police greatly enhanced the capabilities of the surveillance state. Compared with DSP units, regular police possess many advantages, making them useful in carrying out routine surveillance. Most obviously, the frontline police constitute a much larger force. In addition, because the police are embedded in the community—patrolling streets and staffing civilian-facing police infrastructure—they have extensive local contacts that can facilitate intelligence gathering and recruitment of informants. The physical proximity of police to targeted individuals also allows them to maintain closer surveillance and conduct inspections of targets on the orders of DSP units and other security agents.

The total number of police and police stations is not publicly available, but information provided in June 2005 by a senior MPS official is at least suggestive. As of the end of 2004, China apparently had 43,772 police stations, staffed by about 420,000 uniformed police. Of these, 11,492 stations and 202,060 police were located in cities, 19,414 stations and 157,900 police were in urban townships, and 12,866 stations with 60,020 police were based in rural townships. City stations therefore averaged 17.6 police officers, urban township stations 8.1 police officers, and rural township stations just 4.7 officers.[53] But these are not all of the regular police—not by a longshot. A January 2014 MPS news release disclosed that, as of 2013, 556,000 police, only 27.8 percent of the national police force at the time, were assigned to stations.[54]

Public security bureaus and departments also employ "assistant policemen" (*fujing*), who are not counted among uniformed police. These greatly enhance manpower available to staff police stations.

In Guangdong, the provincial PSD recommended in May 2021 that police stations in areas with "complex public order situations" hire two assistant policemen for every uniformed policeman. Stations in other areas were urged to hire one assistant policeman for every uniformed policeman.[55] The provincial PSDs of Guangdong and Hebei provinces also require that the number of policemen assigned to rural stations be, in the aggregate, not less than 40 percent of the total police force in the cities and counties.[56]

When frontline police and assistant police are included, the Chinese surveillance state grows in orders of magnitude as compared to what would be possible with DSP units alone. In fact, police perform most routine surveillance functions, allowing DSP units to focus on high-priority targets and cases.

Police Surveillance Operations

The surveillance responsibilities of regular police are defined in "Protocols for the Duties of Public Security Police Stations," issued by the MPS in 2002.[57] These duties include enforcing the Key Populations program and recruiting informants to provide intelligence relevant to social and political stability and public order. In particular, police are to enlist informants who can speak to the public's reaction to major domestic and international events and issues and who can provide information about activities of hostile groups and illegal organizations. The most useful informants often are those who have contact with the "masses" in the course of their regular working life, or are those who can interact with key individuals without provoking suspicion.

As indicated by the performance metrics of a district PSB in Zhengzhou, the regular police are evaluated in part on the basis of their facility with DSP tasks. Stations are ranked using a hundred-point system, and up to twenty points can be earned through satisfactory performance of DSP duties, which in theory suggests that these responsibilities consume roughly 20 percent of a community police officer's time. The twenty points are further broken down according to a precise rubric. Monitoring KP earns four points. "Maintain[ing] awareness of the activities of Falun Gong, other evil cult groups, and illegal religious groups and organizations" earns two points. Each police officer is required to recruit at least two public-order informants (*zhi'an ermu;* literally, "public-order eyes and ears.") and three other informants each year, netting another two points. Urban stations must "collect twenty pieces of information related to political security protection every quarter," while "rural [stations] must collect fifteen pieces of such information," again gaining two points. And so on.[58]

Activity summaries published in local yearbooks confirm that regular police perform most of the routine operations of the surveillance state. Among these is surveillance of key individuals. For instance the 2002 PSB yearbooks from Beijing and Leshan, in Sichuan Province, reference controlling and managing KI.[59] A late 2010s report from the Yuwangtai District of Kaifeng, Henan, describes police carrying out regular KI verifications and database updates. Elsewhere in Kaifeng, officers had army veterans under constant surveillance.[60] The May 1 station in Weidong District of Pingdingshan, also in Henan Province, reports that in 2019, it kept tabs on the activities of "evil cults," "domestic security KI," persistent petitioners, mentally ill people, and army veterans, often in collaboration with other government entities. (Naming public facilities after May 1, the international Labor Day, is a common practice in China.) Another station in the district claims that, during sensitive periods in 2019, its officers met with KI three times per day. Similar surveillance practices are used in other police jurisdictions.[61]

A widely used police surveillance tactic is known as door-knocking. Most likely this refers to a surprise visit to a target's residence, for purposes of monitoring and intimidation. Usually, door-knocking is conducted jointly by DSP officers and police responsible for the community where the targets reside.[62] In 2016, the Hongyanglou station in Yuwangtai District launched a door-knocking operation against Falun Gong practitioners. Other stations in the district carried out similar operations in 2019.[63] Police in one district of Shenzhen seem to be exceptionally fond of door-knocking. In 2017 they paid 750 visits to sixty-seven KI, averaging a visit per month to each.[64]

Interviews with exiled dissidents and activists who were almost certainly classified as KI when they were in China suggest that DSP officers take the lead in monitoring high-value targets, with regular police assisting. When DSP officers make an initial visit to the home of a target, a local police officer will lead the way and introduce the officers to the target. Thereafter, direct contact between police and the target is limited; contact comes instead from DSP officers.[65] Teng Biao, the human rights lawyer, reports that a community police officer would occasionally call him to ask his plans and whether he was home.[66] However, Hua Ze, who worked with Teng, recalls more active police involvement: a local officer would often visit her aunt, who happened to live in the same apartment building, and ask her to "keep me under control." The officer would make implicit threats to ensure that Hua's aunt complied.[67] During sensitive periods, such as the June 4 anniversary of the Tiananmen crackdown and during the Beijing Olympics in

2008, local police step up their involvement in surveillance of high-value targets. An academic in Beijing reports that during sensitive periods, uniformed officers and police assistants would follow him wherever he went.[68]

If frontline police play varied roles when it comes to interacting with surveillance targets, they are unambiguously active when it comes to recruiting informants.[69] The Hongyanglou police report that, in 2016, they recruited thirty-six spies and informants and used the resulting intelligence to "properly handle" mass incidents. In 2017 they gained forty more recruits. The Xinguanmeng station, also in Yuwangtai District, reported similar accomplishments in 2019. In its 2019 report, the Shenli station in Puyang, Henan Province, describes placing informants around KI and embedding them in unspecified "special groups"—likely religious organizations. Police recruited the KI themselves as informants: officers were tasked with developing "friendships" with KI and thereby "access[ing] inside information . . . to learn about their plans." The 2019 yearbook from Longhui County, Hunan Province, notes that all police at one of its stations utilized spies and informants to collect information about terrorism and stability-related KI.[70]

A final responsibility of frontline police is inspection and control of "battlefield positions"—venues including hotels, internet cafes, rental housing, and printing shops. As I detail in Chapter 6, enforcing restrictive rules on these sites constitutes a pillar of the surveillance state. In 2018, police in Eshan Yi Autonomous County, Yunnan, reported inspections of internet cafes, hotels, rental housing, and entertainment establishments. In 2019, officers of the May 1 station in Weidong District regularly inspected hotels, rental housing, and package-delivery services. The station in Guanlan, Shenzhen, claims that in 2017 it inspected and enforced regulations focusing on internet cafes, hotels, rental housing, and other "complex venues."[71]

The Ministry of State Security

Perhaps the most secretive of all Chinese security agencies is the Ministry of State Security. Official Chinese publications contain little information about either its structure or its activities. The MSS was established in 1983, as the result of the merger of the CCP's Central Investigation Department and the counterintelligence department of the MPS. Because the primary missions of the MSS include covert overseas operations and domestic counterintelligence, its role in domestic surveillance has received little attention. But close examination of yearbooks published by local authorities and universities, as well as interviews with exiled dissidents, reveal that the MSS plays an active, albeit narrow, role in domestic surveillance.

Organization of the MSS and its Local Agencies

Shortly after the establishment of the MSS in 1983, local governments began setting up provincial, municipal, and county MSS departments or bureaus. However, some jurisdictions took more time than others. For example, the state security outfit in Beijing's Shijingshan District was established only in 2005.[72] The naming conventions for local MSS agencies vary; for the sake of simplicity, I will refer to them using one of the common terms: state security bureau, or SSB.

Administratively, local agencies of the MSS report both to a superior MSS agency and to the political-legal committee of the relevant city or county CCP organization. The minister of the MSS is a member of the Central Political-Legal Commission, while directors of the SSBs are also members of local political-legal committees. The principal officers of SSBs are vetted and appointed jointly by the superior state security agency and the local party committee. Operationally, however, SSBs are guided primarily by their superiors within the state security apparatus.[73]

While the structure of the MSS has been described by scholars, little is known about the internal organization of the local agencies.[74] The SSB of Linjiang, Jilin Province, reports that as of 1997 it had a "general office" and a "missions department."[75] Disclosures by the CCP committee within the municipal SSB of Xingtai, Hebei Province, provide more useful clues. The 2002 report indicates that the party had one general branch within the bureau and five sub-branches. Since the party typically has one sub-branch in each operational department, it is reasonable to assume that this particular SSB has five specialized departments. At the time, the Xingtai SSB discussed missions consisting of "special case investigations"; "basic task research"; "development of clandestine forces"—that is, spies; "intelligence," "breaking special cases," and "information work."[76]

SSBs maintain close collaboration with local party committees through the party's PLCs, on which SSB heads serve. Local authorities provide intelligence and information to the SSBs; in turn, SSBs offer local authorities training and technical and operational support with respect to security-related matters.[77] For example, Wuhan has reported that, in 1997, some 370 "national security small groups" inside government agencies, enterprises, and other institutions provided unspecified "assistance" to the municipal SSB. Members of these groups reportedly "paid close attention to a range of factors that could affect social and political stability" on behalf of the municipal SSB. The political-legal committee of Wuhan's Hanjiang District discloses that it provided intelligence and information directly to the municipal SSB.[78]

MSS Surveillance Activities

Local MSS outlets have a broadly defined mission. When, for instance, Xinyang's SSB was established in 1999, its main responsibilities were to conduct "investigations and research on the status of enemies" and to build "clandestine forces." The mission of Linjiang's SSB was to engage in "counterintelligence and other work related to state security" and to wage a "struggle on the covert battle front."[79] How exactly the official mission translates into practice is, however, rarely stated. We must infer from carefully worded reports and from witness statements. These suggest that SSBs perform a wide range of routine law-enforcement operations but are also intimately involved in the activities of the surveillance state.

SSBs sometimes perform conventional law-enforcement functions, most likely in a supporting capacity that utilizes their technical expertise in counterintelligence. For instance, SSBs may assist police by surveilling the communications of criminal suspects. The Dali SSB, in Yunnan Province, has reported collaborating with the SSB in a neighboring prefecture to crack a major drug trafficking case in 1998.[80] The SSB of Ezhou, in Hubei Province, claims to have participated in a "strike-hard" campaign in 1996 by providing unspecified "services" to other law-enforcement agencies. Chengdu's SSB acknowledged its participation in the same 1996 anti-crime campaign.[81] The SSB of Hengyang, Hunan Province, has disclosed that in 2002 it assisted local police and prosecutors in investigating "economic crimes"—an official category of criminality in China, covering smuggling, counterfeiting, fraud, tax evasion, and several other activities.[82]

As a state security agency focused on foreign intelligence, the MSS tasks local outfits with a number of national security–related tasks. One key function of local SSBs is to provide national security briefings for government employees who travel abroad on official business and then debrief them on their return. The briefing is likely intended to prevent disclosure by these government employees of "state secrets"—a designation that is often vague and broad. SSB agents evidently use debriefings to scrutinize travelers' activities abroad and collect whatever valuable information might have been gleaned during their trips. Given the frequency of references to such briefings and debriefings, they appear to be standard practice.[83] Local SSBs also screen foreign investment projects within their jurisdictions in search of national security concerns.[84]

In furtherance of their counterintelligence mission, local SSBs mainly target fulltime residents who are foreign nationals as well as people from Taiwan, Hong Kong, and Macao. The SSB chief in Qufu, Shandong, discloses that, in the late 1990s, his bureau "maintained a timely awareness of the situation of full-time foreign nationals" and individuals from Taiwan, Hong Kong, and Macau. The SSB of Diqing Tibetan Autonomous Prefecture, in Yunnan Province, required that hotels providing lodging for foreigners sign "security responsibility agreements," presumably to facilitate the SSB's monitoring of their guests.

While local SSBs focus on outsiders, they do target some Chinese citizens as well. The same SSB chief in Qufu claims that his bureau cast a wide net, covering people from "various local social strata" and including "key individuals"—a designation applied only to citizens.[85] Religious establishments in ethnic-minority areas may also come under the surveillance of local SSBs. The SSB of Diqing Tibetan Autonomous Prefecture reports that, in 2013, it "conducted operations to maintain surveillance of important monasteries."[86] The SSB of Liangshan Yi Autonomous Prefecture, Sichuan—home to large numbers of Yi people, China's largest ethnic minority—describes targeting "hostile forces in and outside our borders and national separatist elements."[87] A report by the SSB of Kashgar, in heavily Uighur Xinjiang Province, claims that, of the forty-three pieces of intelligence and information it reported to higher authorities in 2000, only five were used by Xinjiang's MSS department, and the rest were used by local party committees and government agencies. This suggests that the bulk of the information collected by the Kashgar SSB was related to ethnic unrest in situ, not espionage by foreigners.[88]

Indeed, judging by the frequency of references to the MSS's role in surveillance of ethnic minorities, the MSS is likely the security agency with primary responsibility for surveillance in minority-heavy areas, especially Tibet and Xinjiang. The obvious explanation is that the Chinese government is convinced that external support is fueling unrest in ethnic minority areas, and the MSS is well equipped to address this threat thanks to its counterintelligence capabilities.

Like DSP units and police, local SSBs conduct surveillance using informants. Most references to such activities in local yearbooks are vague and general. For example, in summarizing its work during the 1996–1998 period, Qufu's SSB states that it "strengthened the building of our intelligence and information network." The SSB of Jimo District, Qingdao, reports that, in 1994, it conducted "multichannel intelligence and information collection."[89] However, in some instances, local SSBs have been more detailed.

For example, Diqing's SSB reported in 2013 that it recruited an unspecified number of "people's defense line liaison personnel" (*renmin fangxian lianluoyuan*) and that some of these were enlisted during security reviews of hotels catering to foreigners. The SSB of Jingdezhen, in Jiangxi, claims that in 1999 it promoted the development of "information gathering" by offering financial incentives and conducting training courses. The Jingdezhen SSB specifically mentions that it provided information from its own sources to the municipal party committee after the crackdown on Falun Gong and the NATO bombing of the Chinese embassy in Belgrade.[90] During the crackdown on Falun Gong in 1999, the SSB in Liupanshui, Guizhou Province, collected "a large amount of internal and early-warning information and intelligence" that "provided support to relevant departments." That is, informants—internal sources—generated information useful to those participating in the crackdown.[91]

The role of local SSBs in preserving domestic stability by repressive means is confirmed by numerous disclosures. For instance: a list of priority information collected by the SSB of Liangshan, the heavily minority prefecture in Sichuan. According to the SSB, in 1999 it prioritized the collection of early-warning intelligence and internal information on the fiftieth anniversary of the PRC; the return of Macau to Chinese sovereignty; Falun Gong; the NATO bombing in Belgrade; Taiwanese President Lee Teng-hui's "two-state" theory, which emphasized Taiwanese independence; the fortieth anniversary of the 1959 Tibetan uprising, in which the fourteenth Dalai Lama fled to India after Chinese troops crushed an anti-China rebellion, killing an estimated tens of thousands of Tibetans; and the tenth anniversary of the June 4 events.[92] For its part, in 2000, the Wuxi SSB claimed to have "collected about a hundred pieces of early-warning and inside information related to social and political stability." Evidently "more than ten such pieces were used by the General Office of the CCP Central Committee and the General Office of the State Council"—not by superior state security outfits, reinforcing the point that the intelligence and information collected by these agencies is frequently more relevant to domestic stability than to counterintelligence concerns.[93] Disclosures by the SSB in Yichang, Jiangxi Province, point to the same conclusion. In 1998, the Yinchang SSB provided thirty-one pieces of information to superior authorities. Nine pieces were used by the provincial state security agency and the MSS, while, collectively, the provincial CCP committee, provincial government, and municipal CCP committee used eleven pieces.[94] In other words, the Yinchang SSB, nominally a counterintelligence outfit, seems to be most valuable as an instrument of stability maintenance.

Local SSBs do not merely process and report information; they also participate directly in information collection through surveillance. And the SSBs take part in security operations that can result from intelligence-gathering. While DSP units have primary responsibility for domestic intelligence and security operations, SSBs are mobilized for stability maintenance during politically sensitive periods and ahead of major holidays, helping to provide intensive surveillance of potential troublemakers. Not surprisingly, local yearbooks frequently refer to SSBs' "stability-maintenance work." The SSB of Liupanshui, Guizhou Province, performed unspecified "security work" in 1999, during the tenth anniversary of June 4, the Two Sessions, and the fiftieth anniversary of the PRC. The SSB in Hengyang, Hunan, conducted similar security operations during the convening of the Sixteenth National Congress of the Chinese Communist Party in 2002. As a large number of local SSB agencies report participating in security operations during sensitive periods and on holidays, this appears to be a widespread practice.[95]

Interviews with prominent exiled dissidents confirm that the MSS is often called upon to perform domestic surveillance at critical moments. Teng Biao, the human rights lawyer, had no contact with MSS agents before the run-up to the Beijing Olympics in August 2008. But in March of that year, he was kidnapped by MSS agents after taking on the legal defense of a well-known activist and publishing an open letter calling on the government to respect human rights.[96] Another occasion for mobilizing MSS domestic repression was the awarding of the 2010 Nobel Peace Prize to the activist and political prisoner Liu Xiaobo. Hua Ze, who was assisting Teng in collecting signatures for an open letter in support of Liu, was kidnapped, detained, and tortured by MSS agents.[97]

As for direct engagement in domestic security operations, in 2002, the municipal SSB of Jingdezhen responded after learning that Christians were organizing a large gathering at an unknown venue within the city. Agents worked with the local PSB to arrest the organizers and disperse the participants. In 1998, the same SSB "successfully prevented an incident that would have impacted social stability." The SSB of Wuhu, in Anhui province, reports that in 2003 it quashed several incidents involving Falun Gong members.[98]

Finally, interviews with exiled dissidents reveal that local SSBs and the MSS play an extensive role in the repression of high-value targets—a task that, on paper, is supposed to belong to DSP units. Yet high-value targets often encounter both DSP and MSS agents. Targeted dissidents describe several noteworthy operational details about the MSS.

While DSP units typically initiate surveillance of high-value targets after they are identified, the MSS will take over when cases involve foreign contacts, including those in Taiwan. A civil society activist who was initially under DSP watch has reported that her case was taken over by her provincial SSB after she returned from a trip to Taiwan in 2011. The SSB agent assigned to watch her was of similar age, experience, and educational background. In an attempt to cultivate ties with the target, the agent would bring her gifts, such as books and candies for her child. MSS and SSB agents are known to be particularly aggressive when their targets meet with Westerners in China. This civil society activist told me that, after traveling to another province to participate in a conference attended by several well-known American academics, she was called repeatedly by MSS agents who insisted on meeting her at the airport on her trip home. She refused their demands.[99]

Another exile, formerly an academic in Beijing, described becoming an MSS target after he was suspected of providing intelligence to a Western government. Initially he was under DSP surveillance; after the MSS became involved, he was subjected to an intensive ten-month investigation. Initially agents interrogated him daily for seven or eight hours; later, the frequency fell to once a month and then every two months. Agents searched his home and confiscated his computer. The academic suspected that agents installed malware on his smartphone: not long after he came under MSS scrutiny, he discovered that his phone was liable to become extremely hot and was using large amounts of data without explanation. MSS agents also followed his movements and tracked closely his contact with Westerners. He was initially barred from foreign travel and required to notify the MSS before taking any trips domestically. Eventually he was allowed to travel abroad but was required to notify the MSS in advance about his itinerary.[100]

Another target was public health and LGBT-rights activist Wan Yanhai. At first he was monitored by Beijing DSP units, but shortly before the 1995 World Conference on Women, he came under the MSS microscope. Like other high-value targets, Wan was subject to intense surveillance measures during sensitive periods, such as the anniversary of June 4. He was detained by the MSS in 2002 for his involvement in publicizing the plight of AIDS sufferers who had contracted the virus as a result of unsafe blood plasma–donation procedures in Henan Province. As standard operating procedure, MSS agents would interrogate Wang each time he met with representatives of Western donors. The MSS would also contact Wan at the beginning of each year to learn about funding he had received from overseas donors and about the activities of his NGO.[101]

A likely explanation for the energetic participation of state security agencies in domestic security is that such agencies are eager to please local party committees and thereby prove their value. The favors they do for local CCP bosses, whose political fortunes often depend on their ability to maintain social stability, can be cashed in later. And the quid pro quo is appreciated, because SSBs can be something of a backwater. Despite the MSS's reputation as a sophisticated and capable spy agency, its organizational capacity and political status at the local level are much weaker than those of the Ministry of Public Security.[102]

ON THE WHOLE, China's system of distributing domestic surveillance among DSP units, frontline police, and the MSS has served the party-state well. The regime has avoided the potential perils of a powerful secret police agency. The formal division of labor among the agencies remains intact even as those agencies routinely interact and overlap, preventing bureaucratic rivalries detrimental to the capabilities of the surveillance state. And redundancy helps to ensure that the whole mission gets done.

The three components of China's distributed-surveillance model complement each other in terms of their strengths and weaknesses. Lean and mean seems to characterize the DSP units, the principal operational arm of the surveillance state. Judging by China's success in containing dissent and social unrest in the post-Tiananmen era, DSP seems to be working. Not only that, DSP is an excellent value. Units are large enough to oversee day-to-day surveillance and carry out special operations—sometimes with the aid of other agencies—but not so large as to threaten the regime. Meanwhile, community police have the manpower to supplement DSP operations, augmenting DSP capacity and enabling elite agents to focus their energies on high-value targets. Yet, as frontline police, they lack the status and sophistication to become an independent power center; their focus on run-of-the-mill law enforcement and public security keeps them out of the political sphere, except when the party wants them involved. In this way, the same police who respond to routine lawbreaking and public safety challenges contribute greatly to the party's security and help to realize its policy agenda.

Finally, the MSS brings a fearsome reputation to the project of state surveillance. Given its counterintelligence mandate, the scope of its domestic operations is limited. But the party has also shrewdly installed it at the local level, incentivizing it to support domestic security and thereby justify its bureaucratic existence. There are potential downsides here: by diverting capacity to domestic security, the MSS could harm its own counterintelligence mission and foster tensions with its more powerful sister agency, the MPS. At the same time, the presence of the MSS in domestic affairs serves as a check on the MPS, containing the latter.

It is difficult to imagine that China could have so effectively addressed the coercive dilemma without distributed surveillance, and it is similarly difficult to imagine how distributed surveillance could function without the Leninist party-state. The supremacy of the party is embedded in the coercive apparatus and reflected in its the primary mission: not to ensure public safety or even defend the state from foreign enemies but to guard the party's political monopoly. Leninist organization enables the most innovative feature of distributed surveillance: its multilayered framework, which weaves together the party and the bureaucracy by installing security chiefs in party committees at all levels, and party committees within the security bureaucracies, again at all levels. This system allows the party to maintain a tight grip on the security apparatus, ensuring that it does not become a danger to its political masters.

At the same time, the distributed surveillance model invented by the party-state has its flaws. Many tasks performed by the surveillance state seem unnecessary: tracking the activities of petitioners, "cultists," and religious organizations arguably achieves little besides assuaging the party's paranoia. These excesses are potentially counterproductive, antagonizing the public and extending the reach of the repressive apparatus without obvious benefits to the party-state. Time will tell if the party has overcorrected in the wake of its 1989 near-death experience. For now, though, the party seems unconcerned about wasted or self-destructive repressive efforts. The successes of the surveillance states outweigh the failures, as the regime's durability attests.

CHAPTER 4

Spies and Informants

The use of informants to gather leads is a conventional law-enforcement tactic around the world. But while police in democracies and dictatorships alike employ informants, their methods differ in crucial ways.[1] In democracies, police use of informants is subject to strict legal limits and administrative oversight. To be sure, law enforcement agencies in democracies sometimes misuse informants for political reasons—the FBI under J. Edgar Hoover may be the most notorious example—but they are supposed to recruit informants for conventional crime-fighting purposes.[2] By contrast, police in dictatorships routinely use informants for both conventional law enforcement and political repression. Because dictators consider political threats more dangerous than crime—and because they see such threats lurking everywhere in society—they tend to recruit a large number of informants for political spying. In Bulgaria, the secret police had as many as 55,000 informants, roughly 0.7 per 100 population, in 1953. By 1989, when the Berlin Wall fall, more than 1 percent of East German citizens were Stasi spies.[3]

There is good reason for democracies to limit the use of informants to conventional law enforcement tasks and for dictatorships to refuse such limits. In the case of democracies, protections of civil liberties preclude investigating people solely on the basis of their political beliefs. And a society shot through with informants will be riven by distrust, eroding the foundations of democracy and market economics, its frequent companion. Dictatorships do not have these concerns. Indeed, social trust is the bedrock of collective action and therefore, in a dictatorship, to be feared. Political spying thus kills two birds with one stone: it identifies potential threats to the regime and it sows distrust among the population.[4]

In dictatorships, only practical considerations constrain the recruitment and deployment of informants. One potential obstacle to recruitment is leverage: the regime needs to convince ordinary citizens to work for the secret police, a morally unpleasant task for most people. Dictatorships lacking the tools of persuasion—promises of money, privileges, career advancement, and protection from (state-imposed) harm, for example—are limited in their capacity to build a large network of informants. A second challenge is operational. The more informants the state relies on, the larger the workforce of secret police it needs in order to supervise and take advantage of those informants. This is a cost many dictatorships cannot afford. A final challenge lies in the productivity of informants. How to get the most out of them, so that that they don't waste their handler's valuable time? Some informants—usually those reluctantly coerced into spying—do as little as possible. Others are eager to demonstrate their value by generating huge quantities of intelligence, much of it junk.

In terms of recruiting and handling spies and informants—again, the formal distinction in the Chinese context is that spies (*teqing*) are recruited and supervised by police, informants (*xinxiyuan*) by political authorities—the CCP's system of distributed surveillance functions with impressive effectiveness. Due to the CCP's extensive presence in the economy and society generally, police and local party organizations have enormous leverage over citizens and can recruit large numbers of them with relative ease.[5] The division of labor among secret police (DSP and PSB units), frontline police, and local party organizations (mainly political-legal committees) allows the secret police to focus on a small group of select spies assigned important missions. The party, drawing on its vast membership, can easily recruit a larger network of informants with more routine responsibilities.

Still, as I discuss below, the party struggles with the productivity problem. China's surveillance state is a clear numerical success: as a share of population, informants in China are as bountiful as they were in the former East Germany. But the quality of their work is suspect. As much as 60 percent of xinxiyuan—the informants, recruited by local political authorities, who perform routine surveillance—apparently provide no intelligence, and only a quarter of the intelligence these informants generate seems to be valuable (see Appendix).

Despite an imperfect network of informants, the party can take some pride in its institutional innovations. The Chinese system is characterized by wide and multilayered coverage. The different categories of spies and informants perform distinct tasks. By maximizing the Leninist regime's organizational presence at the grassroots level and its control of access to resources and opportunities, the party has built and maintained a network of informants that even the Stasi would envy.

Spies (*Teqing*)

The Chinese government began using teqing in the early 1950s.[6] The practice was suspended from 1967 to 1973 due to the Cultural Revolution, but otherwise spying has been an essential tactic of the Chinese regime as it seeks to uncover and suppress the activities of political opponents. In the post-Mao era, the use of teqing has been gradually institutionalized and expanded. In 1981, the Ministry of Public Security (MPS) issued "Interim Rules on the Use of Teqing in Criminal Investigation" to formalize the protocols for recruiting and deploying spies. These protocols were revised and made permanent in 1984, with the issuance of "Detailed Rules on the Use of Teqing in Criminal Investigations."[7] These regulations were again updated in 2001, with a new MPS document, "Rules on the Work of Teqing in Criminal Investigations." In addition, there are interim rules on the use of spies in counternarcotics investigations and inside prisons.[8]

These rules suggest that the purpose of teqing is law enforcement: they assist in criminal investigations. This reflects a CCP propaganda tactic. The Chinese government does not wish to advertise too straightforwardly that police spy on political opponents; instead, teqing are made out to be traditional police informants, seeking to prevent criminal activity and help the police capture lawbreakers. In reality, teqing serve both purposes—they aid in law enforcement, and they report on political enemies.

Although there are formal distinctions between spies and informants, security agencies often conflate these categories in their reports and use a wide range of terminology besides. In addition to teqing, there are liaisons (*lianluoyuan*) and "friends" (*pengyou*). Sometimes spies are referred to explicitly as informants (*xinxiyuan*) or as *ermu*, meaning "eyes and ears"—a term that most often refers to informants. The MPS's 1984 "Detailed Rules," defines teqing as "a generic term used internally by public security and state security agencies to refer to clandestine intelligence-gathering individuals who carry out special missions."[9] In the discussion below, I use the Chinese term teqing, rather than the English spy, both because teqing is a legal category in China and because teqing are not alone in carrying out spying. But teqing are, in some ways, special. Of all the spies or informants recruited by Chinese police, those described as teqing almost certainly are the most select.

Recruitment, Management, and Scale

Recruitment and management of teqing follows strict procedures. Police are not allowed to recruit leaders of criminal gangs, habitual offenders, fraudsters, drug addicts, major suspects of investigations, or relatives of police who handle spies. Only the head of the criminal investigation unit of a county-level (or higher) PSB can approve the recruitment of teqing. Recruitment of those to be deployed outside the Chinese mainland must be approved by a police chief of a county-level PSB and filed with a municipal criminal investigation unit. Provincial authorities must be notified if well-known individuals, senior company executives, or key members of criminal organizations are recruited to work as teqing outside of China's borders. A top-secret file is established for each teqing. The file includes the subject's registration form, photograph, biographical information, an identification number, alias, methods of communication, and code words used for confidential communications with handlers. A teqing's file also tracks his or her performance in past investigations, record of rewards and penalties, the specific intelligence he or she has provided, and actions taken by police investigators in response to that intelligence. Police controllers have access to the files only of those teqing whom they directly oversee.[10]

Once a teqing is recruited, he or she receives training at a secret, police-operated base, which is usually disguised as a commercial entity such as a hotel, shop, or restaurant. Training may last anywhere from a few days to several months. According to a former police officer, training may at times occur in a detention center or prison.[11] The PSB of Sanmen County, Henan, requires that police officers in charge of collecting intelligence organize an annual training session for each of their teqing.[12] Training locations must be confidential, in order to protect the identities of teqing.

References to "officers in charge of spies" (*teguan ganbu* or *guanqing ganjing*) suggest that teqing are managed by designated police handlers. A 2018 work summary from the DSP unit in Baqiao District, Xi'an, and a 2017 summary from the DSP unit in Qihe County, Shandong, confirm that at least some teqing—though probably most, if not all—are exclusively managed by designated officers.[13] In 1996, the Shaanxi PSD assigned 162 officers to control 10,693 spies.[14] Each teqing was evaluated every quarter, and those deemed ineffective were terminated.[15] Qihe County reports that its teqing undergo a rigorous vetting process. The first step is approval of an officer's request to recruit a teqing. Then candidates are interviewed and evaluated. Expenses associated with managing teqing are subject to regular review.[16] And precautions are taken to safeguard identities. In 2003, Shaanxi's PSD claimed to have 112 secret training hubs and meeting places, 43 of which were "commercial bases"—shops operating as police fronts. Profits generated by commercial bases were used to reward teqing and subsidize their operations.[17]

According to an article appearing in a Henan police academy journal, detectives use four methods to recruit teqing. First, spymasters identify, through personal contacts, individuals who are capable of spying. Second, spymasters recruit "average" members of criminal groups known to have committed minor offenses. This provides leverage: these recruits agree to serve as teqing in exchange for lenience. Third, police recruit from among ex-convicts who have completed criminal sentences. Finally, handlers rely on frontline police to identify promising recruits.[18] Because turnover among teqing is high—the average tenure is about two to three years—security agencies, including PSBs and DSP units, plan recruitment in advance. For example, in the mid-2010s the Shandong DSP unit issued a "five-year plan for Domestic Security Protection clandestine forces."[19]

Most teqing are recruited and managed by the criminal investigation units of local PSBs. In fact, local PSBs publicly disclose the specific units charged with recruiting and managing their teqing. For example, the website of Beijing's PSB states that its criminal investigation division is responsible for organizing teqing and operating secret bases for their use. The PSB's "mobile investigation unit," which appears to be an undercover team specialized in fighting street crime, also uses and manages teqing.[20] Although this may seem to suggest a clear distinction between spying to combat crime and spying on political threats, in reality the line is blurred because, as noted above, individual teqing serve both purposes.[21] For example, a teqing may be assigned to watch certain public venues or may be employed in a service job that involves contacts with lots of different people—taxi drivers are useful in this regard. Teqing in either position could fight crime and monitor political threats.

TABLE 4.1
Spies (*Teqing*) Employed by Shaanxi Public Security Department

Year	Number of spies
1957	689
1980	538
1983	1,193
1984	1,907
1985	3,100
1986	2,994
1987	4,383
1988	6,691
1989	9,975
1996	10,693
2001	12,108
2003	14,004

See notes for data source.

As official statistics on teqing are classified, it is impossible to ascertain the total number employed by the police. The only official source that has disclosed useful information on the number of teqing within its jurisdiction is the Shaanxi Public Security Department (Table 4.1). The data suggest that Shaanxi police dramatically expanded their network of teqing after the mid-1980s. By 2003, the province had 14,000 teqing, twenty-six times the number in 1980 and roughly 4 per 10,000 population.[22]

One simple method of estimating the number of teqing in police employ would be to apply to China at large the 2003 Shaanxi ratio: four per 10,000 people. This would place the nationwide figure at 560,000 teqing in 2022. Another method is to multiply the number of DSP officers by the number of spies each is required to recruit. Such an estimate would be quite rough because the number of DSP officers itself can only be estimated and because recruitment quotas vary. For example, in the early 2000s in Xining, a city in Qinghai Province, each detective was required to recruit at least two teqing, and the deputy police chief in charge of criminal investigations was required to directly recruit one to two "elite" (*jianzi*) teqing.[23] The PSB in the Jinshui District of Zhengzhou, Henan, also required detectives to recruit two teqing each year. But police in Daocheng County, in Tibet, reported in 2004 that each detective recruited only one teqing.[24] And in 2010, the PSB of Sanmen County, Henan, required detectives in its criminal investigation unit to recruit at least three teqing per year.[25]

Let us assume that each DSP officer is required to recruit two new teqing each year. China had about 2 million uniformed police officers in 2013, and based on the limited local data available, it seems that the share of police in a county-level DSP unit ranges between 3 and 5 percent of the force.[26] This implies that there are about 60,000–100,000 DSP agents. If our assumption is accurate, then China's DSP units annually recruit about 120,000–200,000 teqing focused on known or potential political threats.

Many of these teqing need to be paid, another challenge for their handlers. The MPS does not have clear, uniform rules and procedures for rewarding teqing.[27] Its instructions merely state, "Rewards should be given to teqing who provide leads for successful investigations; significant rewards should be given [to teqing who help] crack major cases."[28] But, probably due to insufficient funding, public security agencies frequently fail to compensate teqing.[29] Abuse of funds allocated for spying is also common: secrecy and lax financial control make spying operations highly vulnerable to corruption, as police can claim spying-related expenses without actually incurring them.[30] To address the lack of dependable funding and to pay teqing, security agencies come up with creative solutions. For instance, in November 2005, the PSB in Xining, Qinghai Province, established a special crimefighting fund from which it would pay rewards of 200–10,000 yuan (about $30–$1,500 at the time of writing).[31] As a crimefighting fund, the program was widely appealing to donors, including local businesses. Yet, the beneficiaries of the program were not crimefighters but rather teqing whose responsibilities included political spying. In effect, the PSB misappropriated funds in order to pay secret agents who were presented to the public as run-of-the-mill informants. This method is likely to have been adopted elsewhere as well.

Varieties of Political Spying

Teqing are categorized according to tasks they are recruited to perform. Case teqing are recruited to assist in specific investigations. Usually these are "draw-out operations" (*lachulai*) intended to produce evidence against gang leaders, and the recruit is a minor figure in the relevant gang.[32] But case teqing also target political opponents. Presumably, case teqing cease to be useful after the relevant investigation is completed. A second category, *kongzhi* teqing (spies "controlling battlefield positions"), are recruited to watch public spaces or are employed in circumstances that give them access to a large number of targets—think of hotel staff and, again, taxi drivers.[33] In 2001, the PSB of Xi'an recruited 60 teqing and 800 other informants from among the city's taxi drivers. As the city had about 10,700 cabs at that time, roughly one in twelve taxi drivers was working for the police.[34] The third and final category are intelligence teqing, recruited to collect a broad range of information. Disclosures by the Shaanxi PSD suggest that most teqing in the early 2000s were of the kongzhi or intelligence variety, which makes sense because case teqing necessarily are unique, specialized individuals, with close access to particular suspects. In 2003, of the 14,004 teqing employed by the province's police, 1,458 were case teqing (10.4 percent), 5,733 were kongzhi teqing (40.9 percent), and 6,774 were intelligence teqing (48.4 percent). A very small number (39) were full-time police employees and presumably were paid a salary.[35]

Activists targeted by the regime report experiences of being tracked by case teqing. Zhang Lin, a veteran dissident jailed after the Tiananmen crackdown in 1989, started receiving frequent visits from two former "student leaders" after he was released in 1991. It turned out they were reporting his activities to the police. One of the teqing, who was sympathetic to Zhang's plight, admitted what he was doing, explaining that he had only agreed to spy for the police after six months of imprisonment. He also told Zhang that the police had planted listening devices in his home.[36] Another dissident, Yang Zili, recalls that when he and several friends formed a youth group at a Beijing university in 2000, the local SSB recruited a fellow student to infiltrate the group, record members' conversations, and turn over materials to the secret police.[37]

Spying happens across the administrative hierarchy, including at the county level. The number of teqing working at the county level is relatively small, owing to resource constraints and the relatively few political threats identified among smaller, more rural populations. As shown in Table 4.2, most county-level DSP units recruit or retain fewer than fifty teqing per year, though some manage more. Given the strict protocols governing recruitment and use of teqing, local DSP units may prefer to avoid the teqing label, instead recruiting other sorts of informants not subject to rigorous approval processes and monitoring. Perhaps this is why

TABLE 4.2
Spies (*Teqing*) Recruited or Retained by County-Level Domestic Security Protection Units in the 2010s

Year	County	Number of spies
2013	Weng'an, Guizhou	22
2013	Lezhi, Sichuan	27
2014	Weng'an, Guizhou	19
2014	Qihe, Shandong	86
2014	Bayan, Heilongjiang	15[a]
2016	Hanyuan, Sichuan	56
2016	Beichuan Qiang Autonomous County, Sichuan	25
2017	Fuchang, Shaanxi	86[a]
2017	Zhuanglang, Gansu	120[a]
2018	Xi'an Baqiao District, Shaanxi	3
2019	Yunyan District, Guiyang, Guizhou	66[a]
Average		48

a Reports specify "clandestine forces" (*mimi liliang*) rather than spies (*teqing*).
See notes for data sources.

the DSP unit in Wusheng County, Sichuan, claimed that, in 2006 it recruited twenty-two teqing, twenty-five friends, eighty-six liaisons, thirty-six informants, and eleven "DSP special personnel" (*guobao zhuan'gan*). In 2013, the DSP unit of Weng'an County, Guizhou, recruited twenty-two teqing (including two focused on online activity), 161 "safety and security" informants (*anbao xinxiyuan*), ninety-two liaisons, and sixty friends. The Lezhi, Sichuan, DSP unit, and frontline police reportedly retained twenty-seven teqing in 2013; in the prior year, Lezhi County reported that it had 484 "relatively stable informants" and 576 liaisons.[38] This suggests that, while the number of teqing recruited annually may be within the above-estimated range of 120,000–200,000, the total number of informants engaged in political spying is probably much higher.

Local yearbooks have understandably little to say about the operations of teqing. Nevertheless, it is possible to glean some clues about their targets, as well as the type and quantity of intelligence teqing collect. For one thing, it appears that religious groups are prime targets for teqing infiltration. The county DSP unit in Bayan, Heilongjiang, claims that an "elite" teqing deployed in 2005 helped break up four instances of "illegal religious activities" and assisted in three investigations.[39] The DSP unit in Qihe, Shandong, reports that, in 2016, it successfully turned an "evil cult" practitioner into an effective teqing. The PSB of Arun Banner in Inner Mongolia recruited teqing from detained members of a religious organization. Baqiao District, Xi'an, reported in 2019 that its DSP unit relied on teqing to "control and gain information" about a major mosque. One DSP unit, that of Nilka County, Xinjiang, reports having recruited teqing who were close to the targets of an investigation, confirming a key tactic of infiltration.[40]

Much of the time, yearbooks are less clear about what sort of informants are used, so we cannot know for sure whether teqing or other kinds of spies are at work. In Gongliu County, Xinjiang, the DSP unit reported in 2013 that it deployed "clandestine forces" in key villages, mosques, businesses, and around "key individuals." (The same unit also disclosed that it established five "secret bases of operation," suggesting that it had safehouses from which to conduct covert operations.) Similarly, the DSP unit of Fu County, Shaanxi, deployed "clandestine forces" to spy on "key organizations, sectors, units, and groups" in 2017.[41]

Occasionally, local DSP units report on the amount of intelligence and information provided by their teqing. Compared with other sources, teqing generate relatively little intelligence, probably due to their small number and the high standard of quality to which they are subject: information teqing provide must be new, specific, and substantive. Thus, in Midong District, Urumqi, the DSP unit collected 749 pieces of stability-related information in 2010, but only 36 of the "valuable pieces" were provided by "clandestine forces."[42] DSP-managed spies in Luonan County, Shaanxi, produced a total of sixty-five pieces of intelligence in 1998, but only two were "enemy intelligence"—the most valuable type of intelligence, and likely generated by teqing. The DSP unit of Luxi County, Yunnan, collected sixty-one pieces of intelligence in 2003: officers of the DSP unit generated twenty-four pieces, frontline police twenty-three pieces, and teqing only fourteen pieces. In Dazhu County, Sichuan, the relative contribution of teqing was even smaller: less than 5 percent of the total information collected.[43]

Yet if teqing can be credited with providing only a small quantity of information, their role in China's surveillance state is nonetheless significant. What teqing lack in quantity, they appear to make up for by generating the most critical information. Teqing handle the toughest criminal cases and, more importantly for our purposes, the most delicate investigations of political targets. The information generated by political spying is critical to the project of preventive repression and therefore to the larger goals of security and, ultimately, preservation of the CCP's political monopoly.

Eyes and Ears of Law Enforcement (*Ermu*)

Ermu, "eyes and ears," constitute another category of informants formally employed by the Chinese police. Of particular interest here are *zhi'an* ermu—informants concerned specifically with public order and safety. If teqing are elite spies handled by designated police officers, zhi'an ermu are more like tipsters in the rolodex of a beat cop. They are recruited by, and report to, regular community police officers, not detectives and DSP agents. Ermu perform generic surveillance functions, such as reporting suspicious individuals and activities they witness in daily life. Like kongzhi and intelligence teqing, ermu also monitor high-crime areas, businesses thought to be sites of illegal activity, and other public spaces. Because there are a lot of ordinary police, and because the police need to maintain surveillance of many venues, it is fair to speculate that the number of zhi'an ermu far outstrips the number of teqing.

We can gain some key information about ermu from the MPS's rule on police-station file management. According to the MPS, only the head of the station can approve recruitment of ermu. Each ermu has a classified file, similar to that used for teqing. The file contains background and administrative information: the ermu's application to work with police, registration and approval forms, alias, photograph, biography, evaluations, a log of rewards and penalties, communication and payment information, and, in some cases, notice of termination. The file also includes the ermu's intelligence output—summaries of information provided, the ermu's written reports and records of their verbal reports, and verification of the information provided, among other materials.[44]

Ermu recruitment is a four-step process. When an officer identifies a potential recruit—usually someone thought to be well-positioned to monitor people in public, such as a street vendor, sanitation worker, or security guard—the first step is to vet their political loyalties, social ties, and family situation and confirm their access to surveillance targets. Second, if the potential recruit is deemed suitable, the officer should "bond" with the recruit by building a personal connection. The third step is a tryout: the recruit is assigned simple tasks to test their effectiveness. Finally, if the results of the tryout are satisfactory, the recruiting officer must receive formal approval from the station head.[45] Thereafter, the recruiting officer becomes the new ermu's handler, and the two are required to meet at least once a month. No ermu is allowed to establish contact with any other ermu. An ermu's performance, as measured by the quantity and quality of intelligence provided, is reviewed at least annually. Those ermu deemed ineffective may be terminated upon approval of the station head.[46]

Like teqing, ermu receive compensation. One official textbook on policing protocols states, "Outstanding ermu may receive material rewards," which suggests that ermu do not receive regular pay.[47] However, the various police agencies operate differently. In the late 1990s, Tianjin police paid both performance-based awards and a subsidy to zhi'an ermu, funded with the PSB's operational budget and income from a levy on temporary residents.[48]

Books published by presses connected with the MPS lay out a range of roles that ermu play in the surveillance state. They prove to be versatile informants, useful for both preventive repression and conventional law enforcement. Zhi'an ermu gauge the masses' reactions to party policies and major political events, help in uncovering criminality, report suspicious activities, identify individuals likely to commit crimes and violent acts, and collect information about potential mass incidents and other developments that may undermine social stability.[49]

As readers may expect by now, there is no publicly available official count of zhi'an ermu. But it is possible to estimate their number based on indirect information. Consider: in the early

2000s, the PSB of Jinshui District, Zhengzhou, Henan, required that each community police officer assigned to a station recruit at least two ermu per year.[50] If we assume that this requirement reflects the national average, then the question is how police are assigned to stations. As we saw, official 2013 data placed the number of stationed, uniformed police at 556,000, though I have suggested that the number is closer to 800,000, if, as is explicitly stated in some available provincial security documents, 40 percent of the total police force must be assigned to stations. It is probable that some officers do not recruit, regardless of official requirements, so let us be conservative and hazard that only three-quarters of police recruit their own ermu. If between 417,000 and 600,000 police recruit two ermu a year, then the total number recruited in any given year in China is between 834,000 and 1.2 million. Obviously, this is a rough estimate, based on unconfirmable assumptions. But it provides a sense of the scale of the ermu network.

Informants (*Xinxiyuan*)

Alongside teqing and ermu are *xinxiyuan*, a term that can reasonably be translated as "informants." What differentiates these informants is that their recruitment and deployment is a project of local authorities, carried out according to their bureaucratic needs. There are no central mandates or procedures. Sometimes, informants are known to the public, even as they have no place in any centralized government file system. Use of informants is widespread and flexible; they work for one or another agency or state organization, whether a PSB, a PLC-affiliated stability-maintenance office (*weiwen*), the people's militia, universities, or state-owned enterprises. The roles these informants play are similar to that of teqing and ermu; what distinguishes xinxiyuan is their status and number. This difference is important. The discretion allowed state institutions in hiring ad hoc informants is the basis of a massive nationwide network.

Categories of Informants

Informants fall into two categories, depending on the recruiting bureaucracies and the tasks assigned to them. As with ermu, there are zhi'an (public-order) informants, recruited by police stations and investigative units of the local PSBs to provide tips, surveil criminal suspects and activities, and keep an eye on political threats. The second category are DSP, or *guobao*, informants.

Police sometimes sign up zhi'an informants en masse, hoping to take advantage of individuals who, owing to their line of work, are well-placed to monitor certain targets. Police in Jiande County, Hangzhou, have recruited owners of small roadside shops as zhi'an informants because they observe traffic and passersby and because they are more likely to be victims of crime such as robbery and theft.[51] Police in Putuo District, Shanghai, recruited "rider informants" among food- and package-delivery drivers because they could access residential premises. According to the 2020 Putuo yearbook, police recruited 277 rider informants from eight delivery companies. The informants reportedly provided thirty-three "useful leads" in the previous year. The PSB of Zhangshu, Jiangxi, built up its network of zhi'an informants with "owners of small shops along main highways, sanitation workers, workers in recycling centers, jewelry shops, second-hand mobile-phone shops, and taxi drivers."[52]

References to guobao informants date back to the early 1990s. Shulan, a city in Jilin Province, began recruiting guobao informants in 1991 and signed up anywhere from 110 to 360 of them every year through 2002. The DSP unit in Panshi, Jilin, claims to have recruited 325 guobao informants in 1994.[53] The frequency of references to guobao informants increases in the 2000s. In 2005, police in the Sartu District of Daqing, Heilongjiang, reported 573 *zhengbao* (political security) informants, another way of referring to guobao informants. The 2014 yearbook from Karamay County, Xinjiang, indicates 379 guobao recruits the previous year.[54] The 2010 report from the DSP unit of Longshou District, Jinchang, Gansu, explicitly distinguishes the more select teqing (6 recruits) from guobao (130 recruits).[55]

Some local yearbooks offer a glimpse of the type of intelligence guobao informants collect. Sartu District claims it received a total of 696 pieces of information from guobao informants between 1986 and 2005. The bulk, 323 pieces of information, was related to petitioners; eighty pieces were about illegal religious groups and activities; fifteen pieces were about events and groups outside Chinese borders; fifteen were related to unspecific "enemy intelligence," presumably information about the activities of known political threats; and 182 consisted of "intelligence about events with a significant social impact."[56] In Shanxi's Kelan county, the PSB reported that guobao informants generated a total of 224 pieces of information in 2016. Of these, thirty-seven pieces were related to "evil cults" and sixty-two to unspecified weiwen issues (most likely petitions and mass incidents). Thirty-four were classified as "terrorism-related," and ninety-one referred to unspecified issues "related to Domestic Security Protection."[57] Such disclosures suggest that the bulk of the information produced by guobao informants does not concern conventional national or public security threats. Instead it relates to religious activities and ordinary citizens' legitimate acts of protest.

Exiled dissidents testify to the use of like xinxiyuan-type informants, as opposed to teqing and ermu. Xu Youyu, a prominent political philosopher at the Chinese Academy of Social Sciences, reports that the deputy chief of the academy's personnel department kept an eye on his activities. Xu also says that the maintenance crew working for the property manager at his apartment complex worked for the police. They would create pretexts to enter his apartment. For instance after foreign reporters interviewed him at his home, custodians would claim his electricity meter or something else in his apartment was broken and barge in. Another noted scholar at the academy reported that reception staff at the entrance to his apartment complex worked for the police.[58] These informants are all most likely of the xinxiyuan variety, as evidenced by their relatively crude methods. Teqing and ermu are less likely to be detected.

Weiwen Informants

Weiwen xinxiyuan (stability-maintenance informants), is a catchall category for party activists and other volunteers who perform generic surveillance roles. These informants are recruited by local weiwen offices—affiliated with political-legal committees—often in collaboration with police.[59] Based on local yearbooks and notices issued by local governments, it seems that the practice of recruiting weiwen informants began on a relatively small scale in the mid-2000s. For instance, the Lianchao neighborhood in the Pengjiang District of Jiangmeng, Guangdong, set up a pilot program to recruit volunteer informants in 2006. That same year, a neighborhood committee in Zhanggong, an urban district of Ganzhou, Jiangxi, reportedly employed fourteen weiwen informants to report on public opinion. The political-legal committee of Dinghai District, Zhoushan, Zhejiang, held an inaugural training course for weiwen informants in March 2007. At the time, the district, with a population of 374,000, had 151 such informants.[60] Starting in the 2010s, local governments throughout China greatly expanded recruitment of weiwen informants.

Online notices issued by various local governments list the following program features:

(1) Local political-legal committees run weiwen informants, in coordination with police, the 610 Office, and the letters-and-visits bureau, which handles citizen complaints (petitions).

(2) Some jurisdictions set a quota for informants based on population; others do not.

(3) Ideal recruits work jobs in which they can inconspicuously observe people and activities around them. Security guards, residential custodians, taxi drivers, bus drivers, sanitation workers, and parking lot attendants thus make especially desirable informants. Building directors (*loudongzhang*), elected by residents to represent them in discussions with local authorities, are also promising recruits, as they are often aware of residents' activities.

(4) Responsibilities of weiwen informants include collecting and reporting information that may be relevant to social stability, particularly information related to petitions and mass incidents. Such information includes evidence of public opinion concerning government policies.

(5) Weiwen informants report to designated township and neighborhood officials, who then pass reports on to county or district weiwen office or other relevant departments.

(6) Some jurisdictions have formal provisions for evaluating the performance of weiwen informants; others do not. Informants receive financial rewards according to the value of the information they provide.

(7) The identities of weiwen informants generally are kept confidential but may be made public in some jurisdictions.[61]

Disclosures in local yearbooks and other publicly available materials confirm these online descriptions. For example, the 2009 yearbook of Haizhu District, Guangzhou, reports that its government recruited one weiwen informant for every hundred residents, and many of the informants were residential custodians, parking lot attendants, or owners of rental housing.[62] Of the 9,698 informants working for authorities in Nanjing's Qinhuai District in 2016, 6,764 were building directors, 601 worked for property-management companies, 857 were employees of state-owned enterprises and government institutions, and 455 were real estate brokers. Of this vast network, only 287 informants (less than 3 percent) were dedicated to Domestic Security Protection, indicating that huge numbers of people are recruited to perform generic surveillance tasks even as relatively few specialize in political spying. Qinhuai's government-employed informants—not those working directly for the police—helped authorities gain "real-time" information about the activities of "key individuals."

Elsewhere we learn that, in Helan County, Ningxia Hui Autonomous Region, in 2011, informants were paid 50 yuan for each piece of "valuable information" they provided. Daguan County, Yunnan, set "clear responsibilities and evaluation criteria" for its informants, who also received stipends.[63] In 2013, Acheng District, Harbin, established an information and intelligence working group consisting of its weiwen office, the 610 Office, the PSB, and the letters-and-visits bureau. This working group conducted regular training for grassroots informants, and each department was involved in analyzing and reporting the information and intelligence from their informants. The district appropriated 50,000 yuan to reward the informants.[64] In 2013 the government of Lingchuan County, Guangxi, recruited more than a hundred informants who gained access to "key groups" and "key individuals."[65]

Other Informants

Further substantial networks of grassroots informants have developed outside weiwen offices and PLCs. One is associated with the people's militia. According to the Ministry of Defense, China had 8 million militia members in 2011.[66] If just a small percentage of the militia serve as informants, their absolute number could be in the hundreds of thousands. Not surprisingly, local yearbooks contain frequent references to this category of informant. The county government of Tianchang, Anhui, requires each militia unit to appoint informants.[67] Informants recruited from the militia in one district of Taiyuan, Shanxi, have participated in intelligence collection, online surveillance, and propaganda work. Militia members in Zhoushan, Zhejiang, have been trained as informants. The militia in Shanghai's Xuhui District trained its members to "collect intelligence related to terrorism, underground religious groups, Falun Gong, and the use of new information technology."[68]

Local governments also rely on the services of a large number of *hongxiubiao*—Red Armbands, a reference to the red cloth bands they wear on their arms. These volunteers disseminate propaganda and perform routine public-safety tasks, such as patrolling streets and shopping malls and enforcing traffic regulations. As their red armbands imply, they are not clandestine operatives. Yet they do carry out surveillance, watching for suspicious individuals and activities and reporting them to the police. Although not technically informants, hongxiubiao should be considered a peripheral component of the Chinese surveillance state.

Size and Productivity of the Informant Network

It is impossible to know the exact size of the xinxiyuan network. Indeed, there probably are no national data on the subject at all. With various local and provincial governments designating informants inconsistently, to even collect such data would be an ordeal. We can be reasonably sure, however, that every community in China hosts multiple informants of various sorts. Here I attempt to estimate the size of the informant network, including xinxiyuan, not teqing and ermu. As in other cases, I rely on local information and extrapolate from there.

Recruitment requirements vary dramatically across jurisdictions, making for rough estimates on this basis. In Taihe, Jiangxi, the county PSB requires that each police officer recruit between thirty and fifty "intelligence informants." According to its 2018 yearbook, the government of Beichuan Qiang Autonomous County required every large township to have between five and eight informants and liaison personnel, and every small township to have three to five. As of 2012, in the city of Xinmi, Henan, each village or urban community had two weiwen informants.[69]

Some jurisdictions index informant-recruitment quotas to population. In 2012 Yun'an District, in Yunfu, Guangdong, required one informant for every one hundred residents. In the same year, Gao'an County, Jiangxi, mandated one informant per 100–500 households in urban areas and one informant per 100 households in rural areas.[70] Beijing claimed to have recruited 100,000 "safety and stability informants" in 2014, representing about 0.47 percent of the population.[71] In the same year, Beijing's Haidian District had 11,108 informants, about 0.5 percent of the population. The city's Xicheng District reported 17,158 safety and stability informants, about 1.3 percent of the population.[72] More security-conscious jurisdictions may opt for more informants, as the difference between Haidian and Xicheng suggests: Xicheng, a central district of Beijing, hosts Zhongnanhai, the compound housing the offices and residences of top Chinese leaders. Presumably this has something to do with the increased security measures.

Based on the number of informants disclosed in yearbooks from thirty localities, mostly for the 2010s, I estimate that an average network of informants constitutes 1.13 percent of population and a median network 0.73 percent of the population (Appendix Table 1). This implies that between 10.2 million and 15.8 million Chinese citizens are xinxiyuan, alongside an unknown number of militia informants. If only 5 percent of the 8 million militia are informants, that would add 400,000 to the total.

This is an impressively large network. But how productive is it? Here I rely on what information is available concerning the output of xinxiyuan. The data in Appendix Table 2 show that the quantity of intelligence produced by xinxiyuan varies widely across jurisdictions. In a small number of jurisdictions, they appear to be unbelievably productive: Beijing's Shunyi District reports—implausibly—that each of its informants provided about seventeen pieces of intelligence in 2013. Lianyun District, Lianyungang, Jiangsu is another outlier, though less extreme: it reports that each informant provided roughly three pieces of intelligence in 2014. At the bottom end of the spectrum, informants in Pingdingshan, Henan, and Gaoyao, Guangdong, apparently produced just 0.1 pieces of intelligence each in 2019 and 2014, respectively. Although these jurisdictions rank at one extreme, they are not outliers. Many jurisdictions report low productivity rates.

If the two outliers are excluded, then one informant provides, on average, 0.38 pieces of information per year. This implies that roughly 60 percent of informants—and quite possibly more—report no information at all. Still, if 40 percent of xinxiyuan are active, then we can reasonably estimate that at least 4 million ordinary citizens are providing intelligence to local authorities.

Most of the intelligence this army of informants provides isn't directly related to serious, near-term political threats. Disclosures of DSP work show that only a tiny share of intelligence (3 percent) is categorized as enemy intelligence, concerning the activities and intentions of known political threats. Presumably, this is the most sought-after intelligence. Political intelligence—information about public reactions to government policies and major domestic and international events—accounts for 21 percent of information provided by informants. The overwhelming share of information—76 percent—is social intelligence, concerning broader trends among the population (Appendix Table 3).

Thus, broadly speaking, the Chinese informant network chiefly serves as a means of gauging public opinion. To the extent that it does generate enemy intelligence, that effort falls mainly to informants working for DSP units, not to the broad range of citizens recruited by political actors. Most xinxiyuan have no access to dissidents and members of banned religious groups—the sort of people treated as political threats. And, again, there may not be a great many political threats to contain.

As for the quality of the information generated across the spectrum of spies and informants, we can measure it by ascertaining—to the extent possible—how much information is deemed worthy of the attention of superior officials. When informants provide intelligence, their handlers among the police and local committees vet that material before passing it along to higher authorities. Disclosures of yearbooks from nineteen jurisdictions suggest that only about one-quarter of information collected is reported to higher authorities (Appendix Table 4). This suggests that most information reported by China's network of informants is deemed to be unreliable, unusable, or irrelevant.

None of this means, however, that the informant network is failing to do its job. Quite the opposite. The lack of enemy intelligence and actionable information plausibly testifies to some inefficiency but also to China's political stability—stability secured in part by the ever-present reality of surveillance, which keeps the opposition weak and disorganized.

CHINA'S INFORMANT NETWORK IS LARGE AND INTRICATE, with vast numbers reporting to political, law enforcement, and national security authorities operating at every level of the central and local bureaucracies and party organizations. China's DSP units and frontline police may boast more than a million spies and informants assigned to high-value targets. Add millions more xinxiyuan and it is plain to see what truly distinguishes the Chinese surveillance state: its size, redundancy, and reliance on many kinds of more-and-less capable agents to cover the range of political threats and broad issues of social concern. The masses provide the bedrock of distributed surveillance.

In this respect, Mao's notion of the "people's war" remains embedded in the CCP's strategy for its own survival. True, this strategy of deputizing the masses, which Chinese Communists employed against Japanese invaders and against the Nationalists during the civil war, no longer governs military doctrine. But the people's war inspires the surveillance state: instead of granting the secret police a monopoly on intelligence gathering, the ruling party distributes the task of political spying by empowering local organizations and state-affiliated entities to build their own informant networks. Thanks to its penetration of Chinese society and influence on the economy, the party faces little difficulty in weaving such a large spying network. Indeed, the xinxiyuan—the backbone of the party-run network—were assembled quickly, rising after the social unrest of the late 1990s became a serious threat to stability.

Such a large network is not going to be a model of efficiency. Many of China's informants are passive, and most of the information they provide is of little value. Yet this probably does not bother the party very much. After all, even passive informants serve a useful purpose. The very knowledge that one's neighbors, colleagues, and perhaps relatives—not to mention passing acquaintances or the clerk working at a nearby shop—may be working clandestinely for the state encourages caution. When anyone around you might report you to authorities, you will think twice before saying or doing anything that might get you in trouble. Herein lies the true power of the Chinese surveillance state, and it wouldn't be possible without millions of ordinary citizens watching and listening in the course of their daily lives.

CHAPTER 5

Mass Surveillance Programs

The central task of any surveillance state is to monitor people considered threats to regime security or public safety. This is no small challenge. Mass surveillance—tracking potentially millions of people—is a costly, sensitive operation. Indiscriminate spying is expensive and excessively repressive, but a surveillance program covering too few people risks overlooking potential threats. How to determine the small share of the population subject to scrutiny? And what if this share of the population, though small relative to the size of the citizenry as a whole, is still considerable in absolute numbers?

Dictatorships lacking strong organizational capacity or extensive presence at the grassroots level have proven incapable of instituting and maintaining mass surveillance programs. It is relatively easy for even badly organized dictatorships to blacklist large numbers of perceived threats, but to actually monitor them is another matter. Even communist regimes, which traditionally develop large bureaucracies and achieve considerable penetration of society and the economy, find that effort challenging. When Stalin instituted mass surveillance in the 1930s, the system functioned poorly because standards for categorizing targets were confusing, logistical support was lacking at the grassroots level, and information on targets often was out of date.[1]

Given the huge size of China's population, it seems inconceivable that a mass surveillance program targeting even a fraction of the population would be administratively feasible. Yet the CCP has learned to do what other dictatorships have failed at. The party instituted two mass surveillance programs as soon as it seized power in 1949. Although the Key Populations program (KP, *zhongdian renkou*) was an apparent failure at first, owing to the regime's lack of policing resources and self-inflicted economic and political setbacks, the Four Category Elements program did succeed in tracking and restricting the activities of more than 20 million people over three decades. In the post-Mao era, as this chapter shows, the party has not only revived the moribund KP program and turned it into a powerful police-run surveillance initiative covering millions of people but has also developed an even larger mass surveillance program, Key Individuals (KI, *zhongdian renyuan*), that inherits many features of the Four Category Elements and covers millions of additional people not included in the KP program.

Compared with other dictatorships, including former communist regimes, China's mass surveillance programs stand out for the size of the population monitored, their procedural formality (strict rules govern the KP program), and the amount of labor and organizational efforts invested in them. To be sure, China's two mass surveillance programs, in particular KP, have serious flaws. For instance, it might be argued that they focus excessive attention on political threats at the expense of public safety—although this is probably by design and is not necessarily a defect from the CCP's perspective. Whatever their deficits, these two programs have accomplished the party's objective of keeping all conceivable threats under close watch.

Several factors explain the party's relative success in operating mass surveillance programs in the post-Mao era. One is, of course, the growth in resources available for policing. However, the party has not simply thrown money at its problems. It has used its capacities shrewdly, making trade-offs that enable greater focus on high-value targets. To this end, security agents operate multiple surveillance protocols, allowing intensive monitoring of clear and present threats alongside just enough scrutiny of individuals who may emerge as troublemakers in the future. Thus, some targets may be under sustained observation by informants, uniformed and plainclothes police, and technological systems that monitor their whereabouts, communications, and financial transactions in real time. They may be followed by law enforcement authorities, subject to frequent harassment and interrogation, and unable to leave their homes or travel without permission. Meanwhile, less important individuals will simply be listed in databases, their information on hand in case authorities decide to focus on them later.

As the objective of surveilling known or potential threats is to ascertain the danger they pose to regime security, files on these individuals need to be regularly updated and evaluated—but not every person of interest must be under constant, active scrutiny. Thus both the KP and KI programs categorize multiple types of targets and assign them threat levels, which determine how much attention they will receive and in what circumstances. Even threats considered serious may be largely ignored for a time, only to find themselves under careful watch during sensitive moments.

Such discriminating application of surveillance is made possible in part by the party's penetration of Chinese society—the party cells ubiquitous in communities, commercial establishments, and social organizations allow the regime to build and maintain the labor-intensive infrastructure of mass surveillance. The hukou, maintaining files on nearly the entire population, provides a critical institutional foundation. As previously noted, Chinese authorities also have successfully marketed mass surveillance to the public by positioning it as a necessary tool of law enforcement: essentially, the state justifies surveillance as a guarantor of public safety, even as the system is also used to pursue opponents of the regime. Indeed, available data suggest that most KI and KP targets are not political threats. Yet, by ensnaring dissidents alongside run-of-the-mill criminals, these programs perform the essential task of preserving the CCP's political monopoly.

Key Populations

When it was established in 1950, the Key Populations program was a marginal component of the Chinese surveillance state, due to lack of police and administrative resources. However, in the post-Mao period it has been transformed, with huge investments of money, political support, and manpower. Today, along with the Key Individuals program, it comprises the world's most successful mass surveillance infrastructure, tracking subversive elements, criminal suspects, and other potential threats to public and regime safety. The rules for operating the system, issued in 1957, reissued in 1985, and revised in 1998, have never been made officially available. However, the 1998 revision has been leaked online and informs the description below.[2]

Who Are the Key Populations?

There are two broad categories of KP targets. Reflecting the precedence given to political threats, the primary category comprises, according to the 1998 program rules, "individuals suspected of endangering state security." Members of this category are further divided into seven subcategories, which may overlap. They are individuals:

(1) suspected of subversion, separatism, defection, or treason
(2) suspected of participating in unrest, riots, uprisings, or other activities that endanger state security and social stability
(3) suspected of organizing, participating in, or having ties with hostile organizations or of organizing or participating in other organized activities that endanger state security and stability
(4) suspected of participating in banned religious organizations or of engaging in illegal activities under the cover of religion
(5) suspected of willfully sabotaging national unity, resisting state law, and engaging in propaganda and incitement
(6) suspected of engaging in espionage or of stealing, exploring access to, buying, or illegally providing state secrets or information
(7) suspected of other activities that endanger state security.

The second broad category of KP is criminal suspects. These are individuals suspected of serious lawbreaking, such as murder, rape, assault, human trafficking, robbery, theft, arson, organized crime (including prostitution and gambling), manufacture or possession of firearms and explosives, illegal fundraising, and drug trafficking.

In addition to serious criminals and political threats, three additional groups fall under the category of KP: drug users, those considered likely to engage in violent acts due to interpersonal conflicts, and ex-convicts within five years of release from prison.[3]

Designation and Cancellation

Local police stations enforce the KP program and designate its targets. According to the 1998 rules, a police officer first produces the potential designee's file, which contains essential information such as name, gender, address, ethnicity, date of birth, unique ID number, education, profession, biometric data, criminal record, family and social contacts, the category in which the individual is to be designated, proposed methods of monitoring and control, and reasons for designation.[4] This file is then reviewed by the leadership of the relevant police station before it is submitted to the supervising county PSB for approval and from there the municipal or prefectural PSB. KP files are reviewed regularly using a similar multistage procedure, potentially resulting in recategorization—a designee could come under suspicion of some new crime, say—or cancelation of the designation.

There seems to be some variation in the exact contents of KP files. At the very least, we know that PSBs may issue their own rules requiring information beyond that required by the central government. For instance, rules issued by the PSB of Zixin, a city in Hunan Province, require that KP files include certain legal documents, such as court judgments and certificates of release from prison; information and incriminating materials provided by informants; transcripts of interrogations; personal political history; lists of social contacts; and materials concerning the designee's close associates. In addition, when new materials are obtained in the course of routine evaluations, this, too, is to be retained in the KP's permanent file.[5] Other PSBs may have the same requirements, or similar ones.

The Scale of KP Surveillance

Information about the KP program is classified, so it is impossible to obtain official statistics on the number of KP designees. Again, we must turn to yearbooks and gazettes published by local governments and PSBs. Sometimes these reveal relevant information, which we can use to estimate the scale of the KP program.

Information from the 1980s is scarce relative to other periods, most likely because local governments lacked the resources to publish yearbooks. What we do know suggests that, over the course of the decade, an average of 3.5 people per thousand were KP targets (Appendix Table 5). The proportion of designees appears to have grown after 1983, perhaps as a result of more effective implementation of the program. In the 1990s, on average, 4.7 people per thousand were KP designees (Appendix Table 6). A significant source of the increase is likely the strike-hard anticrime campaign launched in 1983, which led to high rates of incarceration and, eventually, large numbers of ex-convicts who were automatically placed in the KP program after their release. The first decade of the twenty-first century saw a dramatic drop in the proportion of KP designees, which fell to 2.7 per thousand people, on average (Appendix Table 7). In the 2010s, the size of the KP program increased again, reaching an average of 3.5 per thousand people (Appendix Table 8).

Still, these shifts are fairly minor; perhaps the most notable feature of the KP program during the post-Mao decades is its consistency. The sheer costliness of the KP program discourages its expansion. Instead, it has been augmented by the larger and less formal KI program. As detailed later in this chapter, KI relies mainly on the resources of local governments and state-controlled entities, rather than those of police. This enables more surveillance, without further taxing the state's policing resources. Then too, overall political stability may be the crucial factor underlying the KP program's steadiness.

Breakdown of KP Targets

The limited information occasionally disclosed by local authorities reveals that, during the post-Mao era, the majority of KP targets have been criminal suspects and public safety threats, not political threats. In 1981, of the 44,622 designated individuals in Heilongjiang, 24,692 were criminal suspects and 18,494 were engaged in activities that endangered law and order. Only 1,436 (3.2 percent of the total) were political suspects. Between 1983 and 1985, the average number of political suspects under Heilongjiang's KP program was 1,833 per year, accounting for about 1.1 percent of KP designees in the province.[6]

The PSB of Changchun, a city in Jilin province, reports that in 1986, there were 3,548 individuals under surveillance in the municipal KP program. Only 145 (4 percent) were "politically dangerous" individuals, while 1,850 were criminal suspects.[7] Likewise five jurisdictions in Zhejiang Province in the 1980s reported that "counterrevolutionaries" constituted just 2 percent of KP targets. About 80 percent were suspected of crimes or were deemed threats to public security.[8]

This trend appears to have persisted. Data from the urban Jianhua District in the city of Qiqihar, Heilongjiang, indicate that political suspects constituted a minuscule percentage of KP targets during the 1990s and early 2000s. Of 2,633 residents designated KP in 1996, only eight were considered politically dangerous. The bulk of the targets were suspected of criminal activities (921) or were released from prison or labor camps (642). The number of political targets rose to fifty in 2001, most likely because Falun Gong practitioners were added to the KP program's political category in 1999. On average, political targets accounted for only 4.4 percent of KP targets in the district between 2001 and 2005.[9] Similarly, Lingbao, Henan, had 627 KP targets in 2000, only 22 of whom were "serious threats to state security or suspects of serious crimes." The rest were ex-convicts released from prison and work programs in the previous five years.[10] Finally, Wuhu County, Anhui Province, reported in 2003 that, of its 1,015 KP targets, only seventeen were suspected of "endangering state security."[11]

If the KP program is largely a tool of conventional law enforcement, then how effective is it in this regard? The police have two key metrics to gauge the program's contribution to law enforcement. The first is the value of information provided by KP targets themselves. Data reported by police gazettes in six Zhejiang counties and cities between the late 1980s and mid-1990s suggests that roughly one in ten KP targets provides leads to the police.[12] The second metric is the proportion of KP designees who become "targets of striking"—that is, targets not just for surveillance but also detention, arrest, or prosecution. The PSBs in the same Zhejiang jurisdictions disclosed that, for the same period, about 5 percent of KP designees became targets of striking. That only a small percentage of KP designees are detained, arrested, or prosecuted can be interpreted in two ways. On the one hand, if the KP program is designed to catch criminals through surveillance, it is of dubious effectiveness because only a small share of designees are nabbed under the program. On the other hand, the KP program may be a powerful deterrent against crime, ensuring that few people under surveillance cause trouble.

As for why such a small share of KP targets are political suspects, a major factor is that few people were convicted of counterrevolutionary crimes or "crimes endangering state security" in the post-Mao era. Another factor is that the rules of the KP program set a relatively high bar for designating political suspects: endangering state security is a serious matter. But if the KP program tends to ensnare few political targets, this does not mean that they escape the watchful eyes of the state. As we will see below, those deemed threats to social stability or the party's authority may well be KI targets.

Limitations of the KP Program

Lack of data makes it hard to evaluate the KP program's effectiveness, but anecdotal evidence points to several limitations.

First, the scope of the KP program has expanded at a faster rate than have policing resources. When drug users were added to the ranks of KP in 1998 and Falun Gong practitioners in 1999, police did not receive commensurate additional funding to handle the hundreds of thousands of new program targets. As a result, KP duty has police agencies stretched.[13] The challenge is that much greater because police responsible for KP are the same officers whose principal task is to enforce the hukou, a labor-intensive job that leaves little time for active surveillance work.[14]

Second, the increased mobility of ordinary Chinese hampers the effectiveness of the KP program. Chinese researchers and police officers report that it is extremely difficult to keep track of KP targets who lack a fixed residence or long-term employment or who do not reside where the hukou says they are registered.[15] In the past, police could rely on neighborhood committees and employers to keep them up to date about KP targets. But greater mobility in residence and employment has rendered neighborhood committees and employers less useful as sources of information.[16]

The KP program is also designed in such a way that police can get away with shirking. Basically, oversight of the KP program happens on paper, so that police may comply with program rules in a formalistic way but without investing much effort in actual surveillance activities. For instance, police can meet their requirements by collecting "static information" about KP targets while not bothering to acquire any "dynamic" awareness of targets' activities. In other words, police can fill out documents full of background on a target and then ignore them in real time. True, police are required to interview KP targets now and then, but these interviews are generally pro forma and yield little valuable information. In light of resource constraints, police tend to commit attention to high-value targets, such as those considered threats to stability and national security, while other types of threats are subject only to as much surveillance as is needed to comply with party-state regulations.[17]

On the whole, then, we might say that the KP program commits relatively little effort to the surveillance of very large numbers of ordinary criminals, criminal suspects, drug users, and ex-convicts, and more effort to surveillance of small numbers of political targets. This may be seen as a sign of inefficiency, but it accurately reflects the CCP's current priorities.

The Key Individuals Program

The KP program includes only those placed under police surveillance through formal administrative procedures. Complementing it is the Key Individuals program, which covers a loosely defined set of targets selected at the discretion of local authorities, including political-legal committees and the police.[18] Whereas KP targets are designated according to rules issued by the MPS, there are no known regulations for classifying individuals as KI targets, nor are there any apparent protocols for placing them under surveillance. Reportedly, the MPS has seven categories of KI targets and maintains a national platform, the MPS System for Management and Control of KI, dedicated to their surveillance.[19] But this does not mean that the MPS directs local security agencies in decisions concerning who belongs in which category. And the national registry is most likely derived from provincial registries.

Operational rules for surveilling KI designees definitely do exist at the local level. One statement, from the Zhejiang Public Security Department in 2010, is revealing in that it lists seven categories of KI, which most likely correspond to those designated by the MPS.[20] They are:

(1) Terrorism-related subjects

(2) Stability-related subjects, defined as political dissidents; Falun Gong practitioners and members of other banned religious groups; advocates of independence for Xinjiang, Tibet, and Taiwan; Japan-related subjects (likely those involved in anti-Japanese protests); and all types of "rights defenders"

(3) Drug-related subjects

(4) Fugitives

(5) Released convicts found guilty of serious crimes

(6) Mentally ill people responsible for disruption of public order and acts of violence

(7) Key petitioners

In considering the seven official categories, there seems to be considerable overlap between KP and KI. For instance, those suspected of involvement in terrorism or narcotics, and ex-convicts with serious criminal records, may qualify as both KP and KI targets.

Local authorities may designate additional categories of KI designees.[21] For example, in 2011, Sichuan police included among KI targets Tibetans and people affected by natural disasters, who were likely to protest against the government's failure to provide timely or adequate relief.[22] In 2004, Benxi, Liaoning Province, had seventeen categories of KI targets under surveillance for DSP purposes, including suspicious long-term foreign residents and "suspicious foreign firms."[23] The DSP unit in Urumqi County, Xinjiang, had twelve categories of KI targets in 2002; in 2006 its counterpart in Midong District, in the city of Urumqi, kept a watchful eye on seventeen categories.[24] Each category of political threats may be further divided into subtypes. In the early 2000s, the PSB in Jiuzhaigou, Sichuan, maintained eighteen categories who were considered targets of DSP units. They were further divided into fifty-seven subtypes.[25]

KI in some respects designates the broad range of surveillance targets not labeled KP. In this regard, we can learn something from an undated document on controlling KP targets, issued by the municipal PSB of Zixin, Hunan. The document lists several categories of surveilled individuals beyond the scope of KP. These include "subjects of attention" (*shixian duixiang*), and "subjects of work" (*gongzuo duixiang*). Subjects of attention have been detained for minor offenses, like gambling. Police maintain files on them to support potential future legal actions.[26] Subjects of work include a wider range of individuals, such as those who have been criminally punished but have not been classified as KP targets. Other broad categories of surveillance targets mentioned in local yearbooks include "stability-related subjects" (*shewen duixiang*), "key subjects for surveillance and control" (*zhongdian guankong duixiang*), and "high-risk groups" (*gaowei renqun*).[27] The proliferation of loose categories suggests that there are no uniform or consistent methods for local authorities to apply when designating surveillance targets beyond the bounds of the KP program.

The Scope of Mass Surveillance

Based on available local data tracking KP designees (Appendix Tables 6–8), the program between the 1990s and the 2010s targeted between 2.7 and 4.7 of every 1,000 people. The median in the 2010s was 2.4 KP targets per 1,000 population; the average was 3.5 per 1,000. Using the median suggests that about 3.4 million people were KP targets in the 2010s; using the average puts the figure at about 5 million.

We can estimate the size of the KI program by comparing KP and KI programs in those jurisdictions that offer data on both. As Appendix Table 9 shows, the number of KI targets exceeds the number of KP targets in most of the fourteen jurisdictions for which we have the necessary information. On average, the number of KI targets is about 155 percent of the number of KP targets; the median is about 115 percent. This suggests that the KI program is between 15 and 55 percent larger than the KP program. If between 3.4 and 5 million people are under surveillance in the KP program, then the KI program likely covers between 3.9 and 7.7 million people.

Altogether, there are probably between 7.3 and 12.7 million Chinese citizens under surveillance, whether as KP or KI targets.

Given how KI targets are described in local yearbooks—repeat petitioners, army veterans, protestors, ethnic minorities—it appears that a considerable share of KI targets are perceived as political threats and agents of social instability. In this respect, KI differs from KP, which mostly designates run-of-the-mill criminals.

Surveillance Tactics

The MPS requires that local police maintain close tabs on KP designees. According to the Rules on Managing the KP, surveillance should be carried out using flexible methods and relying on the "masses." Thus the KP program depends critically on the cooperation of residential organizations, neighborhood committees, activists, and informants. Officers in local police stations responsible for KP surveillance are required to perform most of the formal tasks, such as investigating targets and maintaining information on their identities, aliases, physical characteristics, financial circumstances, and social contacts. Additionally, police must verify information about targets' suspicious activities and report anything important to the criminal units of the relevant PSBs. Nonpolice security agents—for instance, border control officers and officials involved in combatting economic crimes—who come across information involving a KP target must transfer that information to the police stations having jurisdiction over the target.[28]

KP targets are treated differently depending on how threatening they are perceived to be. The rules specify that individuals suspected of posing "major present danger" must be placed under priority control and surveillance. In these cases, police are required to use both overt and covert methods. Overt methods include regular visits with the target. Covert methods consist primarily of assigning informants to keep track of targets' activities. Exactly how police execute these mandates may vary; after all, the sheer diversity of targets necessitates tactical flexibility. Local reports provide hints as to the techniques police use. For instance, according to protocols established by a county PSB in Zhejiang in the late 1980s, covert KP surveillance was to be carried out by at least three people.[29] Other jurisdictions may have similar requirements, but it is not clear whether all do.

A close examination of Zhejiang's public security gazettes yields a small number of datapoints indicative of the share of KP designees who were subject to covert and intensive surveillance in the 1980s. In Jiaojiang County, 19 percent of KP targets in 1985 were subject to covert surveillance. The PSB of Yuhang County claimed in 1989 that 40 percent of its KP targets were subject to "investigation and control," a phrase likely referring to covert control. In the city of Jiande, the PSB placed 19 percent of KP targets under covert surveillance in 1985.[30]

Police reports refer to three distinct levels of KP surveillance. Level-1 refers to "individuals posing a major present danger." Level-1 targets are designated by the PSB at the county or district level, and their surveillance is directly supervised by the head of the local police station or the security department of a government entity. Level-2 and level-3 targets presumably are considered lesser threats and therefore are subject to less strict surveillance. The level of surveillance may be raised or lowered as circumstances demand. In the case of level-1 targets, their residences, workplaces, and other locations they frequent may be monitored.[31]

Level-1 targets appear to account for a small share of KP designees. In Changxing County, Zhejiang, 15 percent of KP designees in 1991 were level-1 targets.[32] In Jiaojiang County, Zhejiang, only

2.5 percent of KP designees in that year were level-1 targets.[33] Although more recent data are not available, it seems reasonable to assume that only a small percentage of KP targets are under intense surveillance because local police simply do not have the resources they would need in order to watch large numbers of people closely. Somewhat confusingly—and reflecting the independence of the assorted security agencies—DSP units use a separate classification system for KI targets. Disclosures by DSP units in various localities indicate a four-level scheme, with level-1 targets the least important and subject to the least intensive surveillance. Level-4 targets are the most important and are subject to the most intensive surveillance regimes.[34]

In recent years, security agencies have adopted technology to facilitate mass surveillance. A 2010 Zhejiang document illustrates how one public security agency has turned to new tools to support tracking and control of KI targets.[35] Zhejiang combines "routine surveillance and control"—a human-centered process, in which police monitor the movements of individuals—with "dynamic monitoring and control," which relies on information contained in national, provincial, and local police databases. If, for instance, a target uses banking services or travels frequently, the information generated by these activities can be cross-checked and matched with data hosted on MPS platforms, helping police gain real-time awareness of the location and activities of the target.

Police in other provinces have adopted similar methods for combining insights from databases and human intelligence.[36] The party-state supports these efforts in part by enabling data collection. For instance, in March 2022 a CCP committee mandated the collection of fingerprints, DNA, voice signatures, digital payment information, and other personal data belonging to KI targets involved in financial scams, apparently to facilitate the surveillance of their activities and movements by technological means.[37] Below I discuss the synergy between labor- and technology-intensive mass surveillance in detail, alongside some other tactical considerations important to the operation of the KP and KI programs.

Labor-Intensive Surveillance

Routine surveillance of KP and KI targets is maintained by teams of local police officers, officials, and informants. In some localities, officials adopt a so-called five-to-one method, whereby a team of five is assigned to each target. For example, in the early 2010s, Anyuan, Jiangxi Province, adopted such a system. Each team reportedly consisted of a township official, police officer, village leader, xinxiyuan informant, and one member of the target's family. The PSB of Yushui, Jiangxi, also forced family members of KI targets to participate in surveillance.[38] A monitoring team deployed in Wei County, Hebei, in 2014 consisted of a county official, township official, village official, "party and government" official, and a police officer. Qihe, Shandong Province, used identical five-person teams as of 2013.[39]

Local authorities report routine use of informants and next-door neighbors to monitor KP and KI targets. Police in Beichuan Qiang Autonomous County, Sichuan, report leaving "daily surveillance of KI" to informants. The PSB in Bayan, Heilongjiang, discusses recruiting KI targets themselves as informants monitoring other targets. In Gongliu, Xinjiang, the county DSP unit planted nearly 300 informants around KI targets and others, as well as in mosques and businesses.[40] Meanwhile, local community organizations such as neighborhood and village committees and workplace security departments designate activists or volunteers to subject targets to "assisted education and reform."[41]

Although KP targets are a police responsibility, the nationwide introduction of grid management in 2013 enabled a new system whereby nonpolice personnel assist in monitoring. Within a given residential grid, several grid attendants are responsible for gathering information, conducting patrols, and maintaining security. In the company of police, grid attendants can enter a resident's home to collect information. In an urban district of Zhengzhou, a city chosen as an experimental site for grid management, a police substation reported that, within a two-month period in 2012, police officers and grid attendants together had closely monitored forty-two KP targets.[42]

Another labor-intensive tactic deployed against KP and KI targets by police and civilians alike is door-knocking. As we saw in Chapter 3, door-knocking is valuable for purposes of both monitoring and intimidation. Thus neighborhood officials in Fucheng, Shenzhen, visited members of banned religious groups at their homes once a month in 2017. That same year, in Ximeng, Yunnan Province, police and local CCP members jointly carried out a door-knocking operation targeting "cultists." Two years later, police in Weixi Lisu Autonomous County, Yunnan, directed a door-knocking campaign against "DSP subjects"—political threats. In 2017 the MPS ordered the PSB of Chaoyang, Liaoning, to enter the residences of nearly all subjects under surveillance.[43]

At this point, readers will not be surprised to learn that the intensity of KI and KP surveillance may be raised during sensitive periods. According to a circular issued by an Anhui police station in 2016, during sensitive periods police officers in charge of surveillance were required to check the "KP management and control system" every day to monitor targets' movements.[44] A police substation in a Guangdong village reported that, in 2020, just before that year's Two Sessions convened, police officers made house calls to mentally ill individuals and KI targets. Police officers spoke to their family members, guardians, and neighbors and sought assistance from these associates in "control[ling] the subjects under surveillance."[45] The 2016 annual report of Nanshan Neighborhood Committee in Chongqing claims that the committee assigned one person to monitor every "key stability-related individual" under its purview. During sensitive periods, these individuals were monitored around the clock to prevent them petitioning, coordinating activities, or assembling illegally.[46]

Technology-Intensive Surveillance

Local yearbooks contain references to the use of unspecified technology to surveil KI targets.[47] The collection of DNA information appears to be one of the more recent applications of technology to trace members of "evil cults." The DSP unit in Weng'an, Guizhou, collected DNA samples from registered practitioners in 2015. Its counterpart in Dejiang, also in Guizhou, reported collecting DNA samples of 135 members of banned religious groups in 2014.[48] Presumably police also monitor the communications of surveillance targets, especially their use of smartphones. Disclosures in yearbooks and official sources indicate that police vigilantly monitor targets' internet use.[49]

Newly installed high-tech systems are making real-time surveillance—long an aspiration of the Chinese police—a reality. Operating from the command and intelligence centers of the PSBs, these systems can alert police about the ongoing movements of surveillance targets by tracking internet and mobile phone use, digital payments, vehicle registrations, and facial-recognition results collected on the platforms of law-enforcement agencies across the country.[50] Police use of mobile phone location technology appears to be widespread. A former political prisoner who gained access to a police officer's notebook when it was accidentally left in his home told me that the notebook contained the code "FF3" in reference to mobile phone location tracking.[51] Such tracking systems are integrated into the central government's Golden Shield and Skynet projects, discussed in Chapter 7.

High-tech tracking and analytics systems have been improving for some time, and a digitized information system for KP management was well developed by 2009. According to *Beijing Informatization Yearbook 2010*, the city's management system at that time provided analytics on the movements and activities of KP targets as well as alerts about their behaviors and analytics-driven censuses of the surveilled population. A separate system, the KP Information Management System, tracks personal information, rental-housing information, and abnormal behavior among KP targets.[52] And the MPS's System for Management and Control of KI is a digital platform. Zhengzhou Railway PSB claimed that, by 2015, it had entered information on about 74,000 KI targets into its Big Data Intelligence Platform to assist in joint operations by the DSP, criminal investigators, and antiterrorism authorities.[53] In the same year, the PSB of a district in Shanwei, Guangdong Province, reported that all of its police stations had begun using an unspecified "information system of controlling key targets," greatly improving surveillance of KI designees. The system generated more than ten thousand advanced warnings about the targets' whereabouts and activities.[54]

Zhongwei, a city in Ningxia Hui Autonomous Region, has disclosed that it deployed a surveillance technology platform in 2019 to monitor KI targets.[55] The city provides few details about the platform's capabilities, but a Chinese company that supplies such technology is more effusive. According to the company, its surveillance platforms enable police and local authorities to monitor targets in real time. Officials provide critical identifying data and information about the targets' "dynamic activities," and the system's algorithms attempt on the basis of this information to determine the target's whereabouts. The system also proposes customized surveillance and control solutions. The company has also referred to electronic bracelets worn by KI targets that facilitate monitoring.[56] This is almost certainly the same platform deployed in Guiyang, the capital of Guizhou. Guiyang's PSB claims that its intelligence center maintained round-the-clock surveillance using alerts provided by the "System of Controlling KI in Real Time." In 2018 the system received a total of over 1.3 million alerts, resulting in arrests of 647 individuals. More than a thousand of the alerts were red-coded—the highest level of alert.[57]

High-Value Targets

China's law-enforcement apparatus devotes enormous resources toward tracking and restricting the activities of high-value targets, such as well-known dissidents and activists. Typically, DSP officers are assigned to maintain contact with high-value targets, although occasionally officers in the local police stations will check on them as well.[58] As previously noted, such contacts consist of regular meetings, "tea," and "meals," during which officers will question the targets about their recent activities.[59]

Alongside tea and meals, DSP and *wenbao* units may resort to more brutal measures of surveillance and control during sensitive periods. Teng Biao, the human rights lawyer, recalls that one year on September 18—the sensitive anniversary of Japan's invasion of Manchuria—wenbao officers drove him from his apartment to his university and sat in on his class for the entire day. During the Tunisian Revolution in 2011, police kidnapped Teng and locked him away for ninety days lest he rile up opposition by pointing to the crowds deposing a dictator in North Africa.[60] Wang Tiancheng, a former faculty member of Peking University, recalls that a DSP unit warned him not to go outside on the days around June 4, the anniversary of the Tiananmen crackdown. DSP agents offered to run errands for him so that he would have no reason to leave his apartment. The DSP unit posted a plainclothes officer disguised as a community security guard in front of his apartment. On the first day of President Bill Clinton's visit to China in 1998, DSP agents knocked on Wang's door to make sure he was home. On the second day, they posted several people with walkie-talkies outside his residence. A police officer was posted at every exit from his residential compound (*xiaoqu*). His every movement outside of his home was followed.[61] Wan Yanhai, an activist, says that, during sensitive periods, DSP agents asked him to ride in their cars if he had to leave his apartment, so that they would not lose sight of him.[62] A former researcher for the Chinese Academy of Social Sciences explains that when the political situation "became tense"—as a result of dissident activities or during sensitive periods—the DSP unit would ask him to "cooperate" and would assign police to follow him even when he went to the grocery store or just out for a walk. During the most sensitive periods, the unit would post guards in front of his apartment building around the clock for up to five days.[63]

When dissident Liu Xiaobo received the Nobel Peace Prize in October 2010, China's surveillance state implemented its most restrictive measures to control the movement of high-value targets. The chief of the Beijing PSB's wenbao unit personally went to liberal political philosopher Xu Youyu's home and supervised his kidnapping and detention in a location far from the city. Hua Ze, the activist who helped collect signatures for an open letter in support of Liu, was kidnapped by MSS agents and turned over to the DSP unit in Xinyu, Jiangxi, where her hukou registration was maintained. She was detained in a hotel for two months and watched by eight police officers working in two shifts. Two female officers slept in her room. Many surveillance targets who were not kidnapped or detained around the time of the Nobel event were visited by the police.[64]

The movements and communications of high-value targets are also monitored by DSP units through wiretapping and the use of informants and spies. Xu Youyu recalls having once received an invitation via telephone to attend an event at the French Embassy—within minutes of that call, he received another from police warning him not to go. Zhang Lin, a veteran dissident, took care to power off his mobile phone to evade police tracking. But one day, after he turned on his mobile phone for just a few minutes at a train station, a group of police officers showed up.[65] DSP units assigned to high-value targets complement high-tech measures with labor-intensive ones. To monitor Wang Qingying, a former political prisoner, a DSP unit not only installed several cameras in front of his home but also stationed several plainclothes officers there, where they would play cards and observe his comings and goings.[66]

The intense efforts and resources the Chinese surveillance state devotes to the monitoring of high-value targets reveal both the strengths and weaknesses of the coercive apparatus. On the one hand, the DSP surveillance system is well resourced, intrusive, and effective in fulfilling its mission of tracking and restricting the activities of high-profile regime threats. On the other hand, devoting so much effort, time, and equipment to surveillance would be impossible if the number of threats were greater. The Chinese regime may be able to sustain such protocols when there are few high-value targets. But if their ranks should grow, it is possible that the regime will not be able to afford this resource-intensive approach.

CHINA'S MASS SURVEILLANCE PROGRAMS are marked by both formidable capabilities and inherent limitations. Driven by paranoia, the party places millions of people under surveillance and frequently updates its surveillance programs to address emerging threats. The party's unmatched organizational capacity enables its system of distributed surveillance, exemplified by the two mass surveillance programs, which cover, according to my estimates, about 0.5 to 0.9 percent of the population. Operationally, the activities and costs of the two programs are distributed among different bureaucracies and groups, such as police, local authorities, civilian volunteers, informants, and even family members of surveillance targets.

Although the party-state has relied mostly on labor-intensive methods, it has also aggressively adopted new technologies to strengthen its capabilities. As a result, law enforcement and local authorities are able to snoop on priority targets, in particular those considered threats to CCP rule. They also can control the physical movements of their targets and monitor those movements in real time.

It is not clear just how much the surveillance system contributes to public safety. The political incentives structuring the decision-making of local authorities and law enforcement officials lead to huge expenditures on a relatively small number of political threats as well as individuals that pose marginal risk to the party, such as petitioners and members of banned religious groups.

Yet, perhaps this is precisely what party leaders want. They may technically require surveillance of drug users and ex-convicts, but it is the political threats that really matter. It is fair to say that, limitations notwithstanding, by the end of the 2010s, China's mass surveillance programs had acquired capabilities that would be the envy of even the most powerful and sophisticated dictatorships in history—capabilities mostly directed at political threats.

CHAPTER 6

Controlling "Battlefield Positions"

If China's surveillance programs are expansions and refinements of longstanding approaches to mass action, the adoption of special surveillance protocols for critical public venues, social institutions, and cyberspace exemplifies the CCP's creativity and adaptability. To be sure, subjecting high-crime areas to greater surveillance is a common law-enforcement tactic globally.[1] But the Chinese approach, known as "controlling battlefield positions" (*zhendi kongzhi*), takes a standard policing method to a new level of urgency and sophistication, helping the party preempt opposition in social settings where resistance is most likely to emerge.

As a regime born of violent struggle, the CCP dictatorship has acquired a fondness for applying military concepts to peacetime governance. Party publications and speeches by senior leaders are filled with the language of "annihilation" (*xiaomie*), "mobilization" (*dongyuan*), and "concentrating forces" (*jizhong bingli*). From the party's perspective, social institutions and cyberspace really are battlefield positions.[2] The term is not used metaphorically.

Controlling battlefield positions has been a key Chinese surveillance protocol since the 1950s. Compared with conventional policing of crime hot spots, China's approach is better organized and more intrusive, labor-intensive, and flexible, applicable equally to universities, commercial sites, and Tibetan monasteries. In the early 2010s, when Tibetan Buddhist monks and nuns were self-immolating to protest Chinese rule and breaking laws by hanging banned portraits of the Dalai Lama, police turned to the battlefield-control model to silence them. The same tactics were successfully applied to universities after the Tiananmen crackdown. And, amid the information revolution, the regime has come to view cyberspace as a virtual battlefield, where authorities and informants must fight tenaciously to control critical positions.

The number of battlefield positions is in the tens of thousands; surveilling them requires enormous human resources. True, technology can save on labor, yet officials and police are still needed to perform duties that cannot be automated. Again, the CCP is far beyond garden-variety dictatorships in this respect, thanks to the organizational infrastructure and mobilization capacity that can only be found in Leninist regimes.

Another operational challenge is to decide what exactly counts as a battlefield position: labeling too many venues or institutions would result in stress to precious policing resources and would make it impossible to maintain enhanced surveillance. The Stasi faced just this difficulty. The East German secret police relied on a "focal-point principle" in targeting high-priority venues, but the effort failed because agents, in an attempt to compete for resources, designated too many focal points and effectively rendered their tactic useless.[3]

This chapter illustrates, through case studies, how the Chinese surveillance state has applied its version of a hot-spot tactic to successfully neutralize threats to the CCP. That success yields two insights. First, it again confirms the effectiveness of distributed surveillance, as a wide range of actors is called upon to control battlefield positions. These actors include police, local authorities, university personnel, business professionals, and specialized agents of the party and its security organs. Second, although technology augments battlefield capabilities, the CCP's Leninist organizational infrastructure and public-enlistment capacity are irreplaceable assets—and the key factors behind the effectiveness of battlefield control.

Battlefield Control in Commercial Establishments

Chinese police place certain types of public venues, such as bus stations, public squares, and businesses (called *tezhong hangye*, "special industries"), under close surveillance for purposes of both law enforcement and social control. In most cases, surveillance of business establishments is designed to deter and solve crimes against property. But surveillance of a subset of these businesses, such as hotels and printing shops, is intended to help authorities track political threats. As usual, it is hard to provide national numbers, but reports from two local police departments in the late 2000s credit surveillance of commercial establishments with enabling successful investigations of ordinary crimes—about 8 percent of crimes in one of the jurisdictions and 15 percent in the other.[4]

Local authorities use flexible criteria to designate special industries. In Wuhan, in 2008, about 6,000 establishments were labeled special industries, including hotels, printing shops, secondhand stores, and auto mechanics. Police in the Shijingshan District of Beijing disclosed that, in 1999, they classified more than a dozen categories of businesses as special industries. In addition to hotels and mechanics, these included locksmiths, recyclers, and a storage facility.[5]

Police control these battlefield positions through numerous methods, some technology-driven and others more labor-intensive.[6] In terms of technology, video cameras are used extensively in the surveillance of special industries. In Deqing County, Zhejiang Province, all entertainment establishments were ordered to install CCTV cameras at their entrances and exits and in main halls and corridors in the early 2000s. And all Chinese hotel operators are required to install reporting systems that transmit information about their guests to police. For example, the PSB of Deqing County set up an "information management system" for hotels in the mid-2000s and assigned personnel to examine hotel-guest registrations every three days. If a Deqing hotel does not report its guest registration information on schedule, a police officer is dispatched to investigate.[7] This is likely standard operating procedure throughout China.

Information management is a central component of the surveillance technology monitoring commercial establishments. "Law and public order enforcement information management systems for special industries" are, essentially, suites of software programs that collect and store information about employees, customers, and transactions and transmit that information to police.[8] In fact there is no single, standardized set of such tools; Chinese technology companies collaborate with the police to develop software that meets diverse surveillance needs.[9] Annual reports of the PSBs reveal the widespread adoption of such systems and their pivotal role in the surveillance of hundreds of thousands of businesses across sectors. In 2011, police in Bengshan District, Bengpu, Anhui, installed an information management system of this kind. In 2019, the PSB of Wuhu County, also in Anhui, installed additional technologies, such as facial recognition and internet ID verification, to enhance information management capabilities.[10]

Surveillance of special industries may generate useful political intelligence. For example, surveillance of delivery services allows police—and secret police—to intercept and inspect materials sent by or mailed to surveillance targets. The 2017 annual report of the PSB in Xianyang, Shaanxi, discloses that the city's express delivery services have worked with the municipal SSB and the state postal service to establish a "long-term effective mechanism for joint inspections of express delivery services."[11]

Police also rely on informants to maintain surveillance of special industries.[12] Local yearbooks demonstrate that police seeking informants favor workers who can collect information or observe customers' activities and movements without arousing suspicion. The PSB in an urban district of Jiangxi Province has stated that, by the end of 2008, it had used informants to achieve "comprehensive control" of the district's taxi companies, motorcycle repair shops, and jewelry makers.[13] Recruiting delivery personnel as informants is also appealing to the police because they can easily enter suspects' homes.[14]

Typically, police press business employees to report suspicious activities or to provide intelligence on a regular basis, thus transforming staff into unpaid informants. Police have significant leverage over business operators and employees. As operator permits must be renewed annually—by the police—few business owners dare resist police demands. Police can further increase their leverage by enlisting other regulatory agencies—such as the Tax

Bureau, the Urban Planning Bureau (which is responsible for zoning), and the State Administration of Industry and Commerce—to reward cooperative businesses and punish the rest.[15] The PSB of Panzhihua, Sichuan, reported in 1995 that tips provided by staff working in special industries helped crack many cases.[16]

Police may also require that businesses employ informants. The PSB of Chongming County, Shanghai, required businesses to appoint dedicated "public safety personnel," who received training and were required to provide information. In the early 2000s, the police chief of Zhabei District, Shanghai, ordered each entertainment complex under his jurisdiction to hire at least two informants.[17]

Surveillance of Tibetan Buddhist Monasteries

Religious sites may not seem like battlefield positions, but to the Chinese government, Tibetan Buddhist monasteries qualify if they are located in areas with sizable Tibetan populations. This is not mere discrimination: opposition to Chinese rule is entrenched in monasteries.[18] To snuff out resistance, the party has launched a comprehensive "reform" program. Judging by official documents, this labor-intensive surveillance program was launched in 2011, when the CCP Committee of the Tibetan Autonomous Region decided to "strengthen and innovate management" of the 1,787 monasteries located there. The most important measure adopted was a restructuring of monastery governance, whereby state officials were appointed to reside in the monasteries and serve as their managers and overseers.[19] The policy was later applied to other areas with a large number of Tibetan Buddhist monasteries, placing the party-state's eyes and ears directly inside monasteries across China.[20]

Monastery Management Committees

Management committees dominated by government officials are the CCP's boots on monastery ground. Committee members are jointly nominated by the United Front Department, the Religious Affairs Bureau, Buddhist associations, and local governments, ensuring the party's control over selections.[21] Many committee members hold significant ranks in the party-state.[22] Such committees are not large, but they exercise considerable influence. In Ganzi Prefecture, a monastery with fewer than 300 monks and nuns is overseen by a seven-member committee; a monastery with between 300 and 500 monks and nuns has a nine-member committee.[23]

The most important committee function is to exercise political control over monks and nuns. Committees have authority to organize and oversee "regular and normal" religious activities, accredit monks and nuns, approve admission to monasteries as well as leaves of absence, and supervise religious education. Broadly, they are responsible for "maintaining stability"—the party's euphemism for preventive repression.[24] As I detail below, these committees enforce restrictions on religious activities and education, impose limits on the physical movement of monks and nuns, surveil monks and nuns who travel abroad, deploy intimidation tactics, recruit and employ informants, and gather intelligence.

Restrictions on Activities and Movements

Monastery managers enforce numerous restrictive rules. Monks and nuns are required to obtain from management committees "certificates" that qualify them for residence in monasteries: Sichuan claims that it issued such certificates to 59,900 monks and nuns in 2015. Religious activities involving multiple monasteries and sects are strictly controlled to curtail networking among monastics and prevent protests.[25] Regulations also govern the scale of religious gatherings. For instance, Ganzi requires that activities involving fewer than 1,000 participants be approved by a county-level Religious Affairs Department and those involving more than 1,000 participants be approved by the prefectural Religious Affairs Department. Other government departments, such as security agencies and traffic regulators, may also have supervisory authority over such activities.[26] Any violations of committee rules regarding religious activities can result in severe penalties, including expulsion from monasteries.

Management committees are careful to monitor the whereabouts of monastics. In 2017, the management committee of Rikusi, in Kangding County of Garzê Tibetan Autonomous Prefecture, recorded all departures of religious personnel from the monastery and prohibited them from engaging in religious activities while they were away.[27] Monks and nuns at Byams Pa Gling in Chamdo may participate in religious activities outside the monastery only with permission from both the management committee and the region's Ethnic Affairs and Religious Affairs Departments. At the same time, monks and nuns from other areas are forbidden from entering Byams Pa Gling to engage in religious activities.[28] In Yajiang County, Ganzi, monastics must report their movements and other activities to their monastery's management committee on a daily basis. Special emphasis is placed on clergy on their return from overseas visits, especially those who have traveled abroad without authorization. For instance, in the late 2010s, officials in Daocheng County, Sichuan, conducted thorough investigations of monastery-affiliated individuals who had returned from unauthorized foreign travel.[29]

Surveilling Monks and Nuns

Effective surveillance of monastics rests on two pillars: collection of personal information and identification of key targets, such as monks and nuns regarded as troublemakers and those associated with monasteries having histories of resistance. Tibet has been collecting and storing electronically reports on all monasteries, monks, and nuns since at least 2012. Kangding, Sichuan, claims it collected "basic information" about the 2,085 monks and nuns within its jurisdiction in 2014. Shiqu County, Garzê Tibetan Autonomous Prefecture, reported collecting and storing "basic information" on 7,150 monks and nuns in 2018.[30]

Monastics designated as KI targets receive particular attention and are placed under special restrictions. For instance, they must obtain permission both to leave and return to their monasteries.[31] Kangding reports that it has developed individualized surveillance protocols to monitor monks and nuns under KI designation.[32] The management committee of Byams Pa Gling reported in 2014 that it "strengthened control over KI" and collected detailed information about targeted monastics and their families. There are no formal criteria for designating monastic KI targets, but it appears that authorities believe any individuals who return from overseas visits, those with passports, and those visiting from other counties and prefectures warrant special notice.[33]

Local yearbooks contain relatively few references to the recruitment and use of informants in monasteries. This may reflect the difficulty of finding willing individuals: Tibetan monks and nuns—the sort of people who would make good monastery informants—are, for the most part, resistant to Chinese rule and loyal to the Dalai Lama. When such references to informants do appear, they tend to lack details, perhaps reflecting the sensitivity of spying on Buddhist religious figures. For example, the management committee of Caodengsi, in Barkam, Sichuan, reported in 2019 that it maintained an "accurate grasp of information" provided by informants. In Shannan Prefecture, Tibet, law-enforcement agencies reportedly "strengthened the development of clandestine forces (*mimi liliang*) in monasteries" in 2010.[34] In other circumstances, officials are more likely to report on the number of informants they deploy, the information these informants obtain, and the benefits they secure. We do know that informants work in monasteries, but we don't know much about the scale and nature of their activities.

Surveillance of monastics is not necessarily clandestine: monastery management personnel make routine, direct contact with targets through personal visits. This practice began as soon as management committees were instituted in the early 2010s. In Lhasa, Tibet, for example, committee members have at times been required to meet frequently with monks and nuns and attempt to befriend them, no doubt for purposes of information-gathering and, if friendship proves elusive, intimidation.[35] Authorities in Kangding report that such visits have been useful in learning about the activities, whereabouts, and psychological states of monks and nuns.[36] Meanwhile Daocheng authorities reported that, in 2019, monastery managers held fifty-eight "educational sessions" with KI targets.

It is likely that, during these personal sessions, officials demand and receive pledges from monastics to toe the government line—an intimidation tactic and a means of creating legal peril for monks and nuns who ruffle feathers. In Kangding, committee members required that the families of monks and nuns sign an "Agreement on Liabilities concerning the Management and Control of Monks and Nuns." In 2018, authorities in Daocheng required that all monks in the county's monasteries sign a "pledge to engage in work to fight against self-immolation."[37]

There appears to be some use of technology in the surveillance of Tibetan Buddhist monasteries. The *Wall Street Journal* reported in July 2021 that more than 180 video cameras and facial recognition systems had been installed to monitor seven Tibetan Buddhist monasteries in Sichuan, according to a Chinese government document.[38] But local yearbooks make infrequent reference to high-tech surveillance of monasteries. In 2011 Shannan prefecture in Tibet claimed that its law enforcement "strengthened the technological detection and control capabilities in monasteries" but provided no details.[39]

We don't know much about technological surveillance or use of informants to monitor monasteries. What information is publicly available focuses instead on regulation and surveillance by management committees. We therefore do know that, at the very least, China's labor-intensive surveillance regime is in operation among Buddhist monastics.

Surveillance on University Campuses

For autocratic rulers, university faculty and students represent a permanent threat and thus require intense surveillance.[40] The leaders of post-Tiananmen China are not exceptional in this regard; they count universities, no less than monasteries, as battlefield positions.

Following the crackdown on the student-led democracy movement in 1989, the party implemented an integrated strategy to reassert dominance over universities.[41] At the end of 1991, the State Education Commission and Ministry of Public Security jointly issued a circular on strengthening security in institutions of higher education. Universities and colleges were ordered to improve their security departments' investigative capabilities, in consultation with public security agencies. Authorities also ordered universities to make "full use of the role of administrative departments, labor unions, the Communist Youth League, student unions, and activists among faculty and students" and to "rely on them to gain relevant information about KI and positions."[42]

In February 1997 the State Education Commission and Ministry of Public Security again joined forces to issue new rules. Now they emphasized "mobilization of the masses" to collect information and guard against "infiltration, instigation, and sabotage by domestic and foreign hostile forces, illegal religious forces, and national splittist forces" operating on university campuses. The officials also called on universities to "assist state security and public security agencies to bring an end to activities that endanger state security" and ordered campus administrators to scrutinize student organizations and extracurricular activities. In addition, at this point universities were required to establish systems to "manage" foreign teachers and students.[43] Further rules were added in 2011, when the party set forth specific measures to prevent religious groups from building any presence or influence on campuses.[44]

University authorities have complied by mandating the establishment of files on all faculty and students who are religious believers, strictly regulating approval of academic activities and student organizations, and vigilantly examining foreign NGOs and use of foreign funding. In addition, campus officials vet textbooks and online instructional resources, closely monitor foreign faculty and students, recruit student informants, and regularly report information to law enforcement. In the late 2010s, universities added provisions to safeguard "political security," most likely in compliance with orders from higher party-state authorities.[45]

Organizationally, universities have built a security apparatus that incorporates features of political-legal committees and police agencies. The security department of Shandong University—which, in 2020, had more than 70,000 full-time students—boasts a 610 Office, a cyber section, a "comprehensive management section," and a contingent of security guards.[46] Nankai University's 2013 yearbook indicates that its security department had six sections, including one for "political security."[47]

Targets of Campus Surveillance

Surveillance programs on university campuses mainly target three groups of individuals: KI designees; ethnic-minority students from Xinjiang, Tibet, and Inner Mongolia; and foreign professors and students. It is notable that surveillance of ethnic-minority students began as early as 2000, many years before the onset of large-scale unrest in Tibet and Xinjiang.[48]

University yearbooks contain frequent references to surveillance of KI targets, a category that may include dissident academics and religious practitioners. At the Chinese University of Law and Politics in Beijing, authorities undertook "comprehensive work to educate and manage key groups . . . so as to control and reduce the negative influence and real harm of all key groups."[49] University security officials also describe curbs on proselytizing and apprehension of individuals engaging in such activities. For example, in 2010 the security department of Jiangnan University intervened in three instances of proselytizing. The security department at Nankai University has stated that, in 2013, it selected as "key targets" a small number of faculty and students who were deeply involved in illegal religious activities and assigned dedicated personnel to keep tabs on their "thoughts and activities."[50] That security personnel are able to intervene in proselytizing as it unfolds may be an indicator of the overall effectiveness of the surveillance program.

Ethnic-minority students receive a good deal of attention in university yearbooks. According to Lanzhou University, in the mid-2010s campus security officials assisted state security agencies in investigating and surveilling all Uighur students at the school. Meanwhile the provincial department of education in Zhejiang Province reveals that, in August 2014, it ordered universities to carry out comprehensive vetting of Uighur students, strengthen "management of key subjects," uncover and shut down underground prayer sites for Uighur students, and enforce "management" of Uighur students' passports. In its 2018 yearbook, Jiangnan University reports assisting the Wuxi municipal PSB in "educating and managing" ethnic-minority students.[51]

The intensity of campus surveillance usually increases after incidents of ethnic unrest. After students demonstrated against environmental degradation in Inner Mongolia in May 2011, Hubei's provincial higher-education commission issued an urgent directive to step up security measures. In August 2011, after a series of violent attacks in Xinjiang, the Ministry of Education instructed all universities with a substantial student population from the region to assess any security risks they pose.[52]

Many Chinese universities report on their surveillance of foreign teachers, students, and NGOs. Guizhou University openly acknowledges in its 2014 yearbook that its security department conducted operations "to prevent religious infiltration by foreign students." In its 2014 yearbook, Huaqiao University claims that it successfully prevented "infiltration and sabotage" by unnamed foreign NGOs.[53] Nankai University reports that it "established files on foreign students . . . and built a mechanism to manage them and to monitor their whereabouts."[54] Hefei University appears to keep a particularly watchful eye on foreign professors and students. Its security department in 2010 "intensified the investigation of foreign-related activities and foreign teachers and students and expanded the channels of communication with (superior) foreign affairs departments." In 2011 the university conducted several rounds of investigations of foreign teachers, students, and student organizations.[55]

Univeristy Collaboration with Government Agents

University security departments work closely with state security agencies and with local public security units, mainly DSP and wenbao units. Lanzhou University's 2014 and 2015 yearbooks claim that, besides assisting state security agencies in monitoring foreigners and Uighur students, the university's security department also helped state agents "visit and talk to" faculty members who had returned from studying abroad.[56] Ningxia University reports that it "assisted superior public security, state security, and education departments in their investigations on campus." The security department at Dalian University of Science and Technology claims that it collaborated with local state security agencies to address threats related to Xinjiang.[57] Guizhou University's security department reports having cooperated with the provincial SSB to investigate illegal religious activities on campus and dismantle the organizations behind them.[58] The security department at Nankai University acknowledges that, in 2009 alone, it assisted authorities on forty occasions when they visited the university to "read files, conduct investigations, and pay follow-up visits."[59] The prestigious China University of Political Science and Law in Beijing claims that, in 2011, its security department "paid close attention to the information provided by the public security and state security agencies" and "assisted related departments in operations so as to ensure control over incidents related to political stability."[60]

The regular provision of information and intelligence to security agencies and political authorities became a well-established university practice in the early 1990s. At Wuxi Light Industrial University, which later became Jiangnan University, the security department reported supplying more than 300 pieces of information to its "superior leadership" in both 1993 and 1995.[61] The 2007 annual report from Hefei University of Technology acknowledges that its security department provided forty-one pieces of information to "public security and state security departments." In its 2008 annual report, the university claims to have passed on fifty pieces of information to "the university party committee" in addition to the local PSB and state security departments.[62] In 2005, the security department of Huazhong University of Science and Technology did such an outstanding job of reporting on students, faculty, and staff that it received the designation of "exemplary unit for information work" from the Wuhan SSB.[63]

Universities provide two types of information and intelligence: public opinion and "stability-related" information. The first category includes campus reactions to key events and domestic and foreign policy issues. For example, Jiangnan University's 2001 yearbook describes collecting and reporting student reactions to the Two Sessions and major foreign policy developments. The category of stability-related information is probably very broad.

Jiangnan University reports passing along to "superior authorities" information about the suicide of a graduate student, fights between students, and student strikes over bad cafeteria food.[64] In its 2001 yearbook, the Hefei University security department reports having informed on a "University Alliance in the Hefei area"—presumably an association of area college students whom campus officials saw as a potential threat to the party.[65]

Use of Informants

Effective surveillance of millions of university students and faculty is impossible without informants. Fortunately for Chinese university authorities, recruiting informants from vast student populations is no great challenge. As the party controls access to educational and career opportunities, students are especially susceptible to recruitment: in exchange for spying on behalf of the government, they may gain admission to the party—which improves their chances of getting good jobs—or slots in coveted graduate programs. Although it is impossible to estimate the number of student informants, it appears that all universities employ them. References to regulations of on-campus informants can be found in documents issued by at least two provincial-level education departments. In 2013, Hunan's Department of Education issued "Interim Rules on the Reporting of Information Related to Stability Maintenance and Public Safety in Hunan's Education System." A similar document was promulgated in Hubei province in 2016.[66]

The use of on-campus informants is widely acknowledged; information about their tasks, recruitment, and operations is openly available on school websites. These sites indicate that informants collect and report on individuals and organizations suspected of engaging in subversive activities such as petitions, illegal assemblies, and "reactionary propaganda." Finally, informants report on campus safety and student and faculty reactions to important political events and government policies.[67]

Most informants on campus are students, although some are party officials and administrative staff. South China Agricultural University requires at least one student informant per academic division. At the prestigious Beijing Foreign Studies University, an "opinion informant" is recruited for every classroom.[68] Changsha Medical University and Hunan Institute of Technology both require one to three informants in each classroom. These informants should be students who are party members, or else they may be political guidance counselors—full-time university employees who carry out "ideological education"—or party branch secretaries and deputy secretaries. Informants must be politically reliable and operationally effective.[69] At Hubei University of Economics and the University of South China in Hunan, student informants are recruited mainly from among the secretaries of the Communist Youth League and from among student leaders.[70]

Student informants typically do not interact directly with public security agencies. Indeed, rules issued by universities make it clear that party officials, not public security agents, are in charge of the work of the informants.[71] Thus party operatives handle recruitment, training, supervision, and evaluation of student informants. Student informants do interact with campus security. Informants are encouraged to remain close to campus locations where mass incidents or suspicious political activities are most likely to take place. They are directed to "mingle closely with fellow students and gain a timely awareness of their (fellow students') ideological states of mind."[72]

Operationally, informants are expected to make regular reports to their handlers. The frequency of contact varies across institutions. At Hunan Institute of Technology, informants responsible for collegiate divisions are supposed to provide a monthly report, whereas informants responsible for classrooms must report every other week.[73] Changsha Medical University requires a weekly report from all informants responsible for classrooms, a biweekly report from those responsible for each cohort, and a monthly report from those responsible for departments. Beijing Foreign Studies University requires a monthly report from the "liaison" who supervises student informants.[74] In the case of a critical development or emergency, informants must immediately report any information or intelligence they have collected. Some universities require strict confidentiality about the work performed by informants, but this is not always the case.[75]

The duration of service by informants is not specified in most university regulations. In all likelihood, informants in most universities receive an initial one-year appointment and can expect reappointment contingent on their performance. New recruits constantly replenish the ranks when older informants graduate.[76] Universities offer political and material incentives to student informants. Beijing Foreign Studies University pays each informant "a certain amount of subsidy" per semester. South China Agricultural University gives unspecified "recognition and rewards" to "outstanding informants and information-work activists" and awards them extra academic credit. At Changsha Medical University and Hunan Institute of Technology, informants can expect preferential treatment when they apply for CCP membership and compete for awards. Hubei University of Economics provides material rewards to informants according to the value of the information they provide.[77]

Some universities treat informing as a form of work-study, thereby providing an avenue to compensate informants.[78]

The Cyber Battlefield

Research has revealed a good deal about internet censorship in China—which activities may be prevented, which resources made inaccessible, and so on. Yet, while we know a fair amount of user experiences, little has been written about the actual mechanisms and tactics of online censorship and surveillance.[79] Here I seek to fill in this gap, exploring the "front end" of police operations online.

The party-state has adopted battlefield-control tactics online in order to identify users and track their activities. The government has opted for a two-track approach, dividing surveillance between a party organ—the CCP Office of the Central Cyberspace Affairs Commission—and a new police unit. This arrangement makes clear that controlling battlefield positions online is both a political and a law-enforcement task. Party censors determine which content is to be blocked or filtered, while the police carry out enforcement actions, such as inspecting internet cafes for compliance with regulations, installing surveillance hardware, blocking and filtering the suspects' communications, and conducting investigations and arrests.

The CCP began to assert control over the internet in the mid-1990s, but it did not build a nationally integrated, stand-alone bureaucracy until 2014, with the establishment of the Office of the Central Cyber Security and Informatization Leading Group.[80] This agency was charged with both regulatory and censorship responsibilities. Thereafter, local jurisdictions set up equivalent offices attached to their CCP committees. In 2018, Xi Jinping promoted this leading group to the status of a central commission, whose routine functions are performed by the Office of the Central Cyber Affairs Commission. Recall that commissions are technically party bodies, not government ones. The commission's bureaucratic companion, which shares its mission and personnel, is the Cyber Administration of China. In Chinese it is known as *zhongyang wangxinban*. I'll use the term "cyber agency" to refer to the party-state body that handles online surveillance, as compared to the cyber police—the security agency that shares this task.

Local outlets of the Cyber Affairs Commission are under the umbrella of CCP committees. Shaanxi's provincial cyber agency was authorized to employ up to sixty-one staff when it was formed in 2014. In the city of Chenzhou, Hunan, the municipal cyber agency had thirteen staff in 2017. A typical county-level cyber agency had only three or four full-time staff in the late 2010s.[81] Due to their small size, local cyber agencies lack the workforce as well as the technological capabilities to conduct sophisticated surveillance. Notably, few local yearbooks say much about the technological capabilities of the cyber agencies, even as these reports provide abundant information about accomplishments in terms of censorship and spreading disinformation.

Indeed, routine censorship and promulgating disinformation are the main tasks of subprovincial cyber agencies. The municipal cyber agency of Longnan Prefecture reports that, by the late 2010s, it had used big data and cloud computing to monitor online public opinion and had established a database of essential information. The agency defined 180 keywords and twelve topics that garnered special attention. In 2019, the agency monitored 515,000 pieces of online information about Longnan, of which 8,000 were deemed to be negative.[82] Local cyber agencies also recruit "net commentators" to conduct online campaigns to manipulate public opinion and spread disinformation.[83]

Cyber Police

A special cyber police unit takes charge of internet-related enforcement operations and surveillance of user activities, as distinct from censorship and disinformation. Cyber police, known as "public information network security supervision sections," were first organized in PSBs throughout the country in the early 2000s. Beijing's municipal PSB established its cyber unit in 2000 and staffed it with fourteen police officers.[84] Shandong's provincial cyber unit was established in 2003, and within two years all local PSBs in the province had formed equivalent units.[85] A Beijing public security yearbook reports the completion of a municipal center for "monitoring and controlling public network security" in 2001, seemingly confirming that cyber surveillance systems are housed within the public security infrastructure, not in local outlets of the Cyber Affairs Commission.[86] Yan'an's cyber police unit reports that its main missions include "monitoring and controlling harmful information; collecting, analyzing, and reporting developments on the internet; enforcing regulations on internet cafes; and investigating and dealing with cyber-crimes."[87] Despite the importance of these tasks, the cyber police units in the local PSBs are relatively small. A typical county cyber police unit has about five to six officers. For example, the cyber unit of Shucheng County, Anhui Province, had only five officers in 2020.[88] Its counterpart in Tancheng County, Shandong Province, had seven officers as of 2011. Gejiu County, Yunnan, had five officers in its cyber unit in 2016.[89]

Both the local agencies and cyber police units "patrol" the internet 24/7. Both deploy hi-tech solutions. For example, the cyber agency of Santai County, Sichuan, used web-monitoring technology called Real Time eXchange to maintain constant watch over the internet. Cyber police officers are required to report important developments to the PSB leadership and also to the county party committee and government.[90] Although local yearbooks do not specify which bureaucracy actually performs the task of deleting and blocking online content, apparently the cyber agencies make the determination and then instruct police to execute it. This division of labor is confirmed by the cyber unit of the municipal PSB of Ergun, in Inner Mongolia, which states that it is responsible for "organizing and implementing the 'routine work' of Ergun's cyber agency."[91] "Routine work" almost certainly refers to censoring online content. Reports of cyber police taking bribes from businesspeople to delete certain critical posts also indicate that it is cyber police, not the cyber agencies, that perform the actual task of deletion. In the meantime, there are no press reports of similar scandals involving officials in the local cyber agencies.[92] And when cyber agencies discover online materials requiring investigation, they typically refer matters to cyber police. For instance, as soon as the cyber agency of Yunlian County, Sichuan, became aware of what its 2018 yearbook calls a serious "internet rumor," it contacted the cyber unit of the county PSB to investigate.[93]

This division of labor makes administrative sense. It is unnecessary to duplicate technological capacities across agencies, and to house such capacities within the offices of local Cyber Administration would be a security risk because the commissions are located in government buildings rather than police buildings, which are better guarded. Meanwhile Chinese telecom companies are unlikely to be involved in this censorship process because they lack the requisite political status, security, and law-enforcement authority. It is also police, not the Cyber Administration, that operates China's Public Information Network Security Surveillance and Control System—the so-called Great Firewall. We know this because, for instance, the Tianjin PSB confirms that one of the principal tasks of its work on the Golden Shield project—of which the Great Firewall was a component—was to "discover . . . and appropriately dispose of . . . harmful information online."[94]

Controlling the Cyber Battlefield

The mission of filtering "harmful" content consumes a good deal of attention and energy: enforcing surveillance is labor-intensive, as cyber police must conduct in-person investigations, visiting individuals suspected of violations such as posting censored materials. In 2016, the district cyber police in Baiyun, Guiyang, investigated eighty-five individuals in person. The cyber police in a neighboring district, Yunyan, were even more aggressive, claiming 200 such investigations during the same year.[95] Penalties for harmful online activities include detention, fines, and "criticism and education."[96]

Deterring online dissent is impossible without identifying actual users, so the strategy to dominate the cyber battlefield positions critically rests on detecting the identities of violators and potential violators. The Chinese government uses several tactics to identify users.[97] One simple approach is to trace IP addresses—the unique identification associated with a local network that connects to the broader internet. This is easy to do, as online access is provided by state-owned telecom companies. But more sophisticated users can avoid this sort of identification by routing their online traffic onto a virtual private network. And additional measures are needed in order to identify the owners of social media accounts, which allow anonymous posting. In response, in February 2015, the Cyberspace Administration of China (the bureaucratic companion to the Central Cyberspace Affairs Commission, which technically is a party body rather than a government one) mandated that all internet users provide their real names when registering accounts on chat rooms, the popular app WeChat, the extremely active microblogging service Weibo, and all other social media outlets.[98]

Cyber police surveil access points such as internet cafes and public Wi-Fi networks in hotels, shopping malls, airports, and other venues to identify users operating beyond their home networks. Regulations governing private internet cafes issued in April 2001 mandate retention of customer information, including identities and online activities, for sixty days. Internet café operators must obtain a license from the local PSB and Cultural Bureau, a government agency regulating the entertainment sector.[99] More recently implemented rules require that internet cafés install ID card readers that automatically capture information about all customers for storage in a café-specific database. Customers can gain online access only after scanning their "second-generation" IDs, which store identifying information, including a color headshot. Cyber police enforce these rules strictly through frequent inspections of internet cafés.[100] Police also train "security attendants" at internet cafés, who presumably ensure compliance with regulations and also perhaps spy on customers. In Shandong, the local police force takes credit for training 3,000 such attendants in 2004.[101]

Some jurisdictions even require that internet cafés install video cameras to monitor customers. Local yearbooks show that surveillance cameras were installed in the late 2000s.[102] In its 2014 report, the Baiyun cyber police claimed they began to require that internet cafés install "safety technological equipment" that triggered video recording when "internet KI" scanned their IDs. This system alerts the police and transfers relevant video to law enforcement in real time.[103] In fact, this system was already in operation in some jurisdictions, such as Chengdu, in 2012. A cyber policing textbook explains that when a customer used the ID of a KI target to get online at a Chengdu internet café, the cyber police received an instant alert and could identify which computer in the café was being used. In this case, two police officers were sent to the café to investigate.[104]

As for monitoring public Wi-Fi networks, in the late 2000s, local cyber police began requiring that operators install unspecified "security technical measures."[105] A concerted national effort on this front likely began in 2014. In that year, Wuhan police, for example, initiated a three-year program to install "security management systems" in all public Wi-Fi networks.[106] In 2016 the PSB in Yunyan District, Guiyang, installed 560 public Wi-Fi monitoring systems. Thousands of similar systems were installed in two counties in Sichuan in 2017–2018.[107] On average, a Wi-Fi surveillance device costs 2,200 yuan (about $310 as of this writing), indicating that this surveillance program—including post-installation maintenance and operation costs—requires significant resources.[108]

Cyber police pay special attention to online KI targets. These likely overlap with political dissidents, liberal scholars, human rights activists, members of illegal religious organizations, and practitioners of Falun Gong and other "evil cults." Online KI targets also include some well-known pro-government personalities, indicating the party's paranoia about individuals with a significant public following regardless of their political loyalty.[109]

There is wide variation across China in the number of online KI targets, demonstrating that jurisdictions have wide latitude in making KI designations. In 2018, Hengyang County, Hunan, had a hundred "internet KI" under surveillance and control. The cyber police of Oroqen Banner in Inner Mongolia claim to have had twenty-five KI targets under online surveillance in 2015. Yunhe District of Changzhou, Hebei, reports that its police had in-person contact with sixty-two "internet KI" in 2016.[110] In several jurisdictions, however, the number of online KI targets is much larger. Jishan County, Shanxi, had 1,141 KI targets, roughly 0.3 percent of the county's population, under online surveillance in 2018. Between 2011 and 2014, cyber police in Tancheng, Shandong, "registered and controlled" 3,475 internet KI targets, about 0.4 percent of the local population. Gejiu County, Yunnan, had 562 online KI targets in 2012, about 0.14 percent of the population.[111]

Details of online KI surveillance and control are sparse. At a minimum, however, it seems that cyber police have special files on their targets, as reported in several jurisdictions.[112] It is also reasonable to assume that the online activities of KI targets are closely monitored and that their email and social media accounts are compromised.

A document issued by the PSB of Neijiang, Sichuan, in February 2011 reveals a few crucial details about online surveillance of KI targets. According to the report, the PSB cyber unit was instructed to collect basic information on all types of KI targets, to designate officers to use "various technical means" to scrutinize these individuals, and to use an unidentified special police database to ascertain their online identities. The report mentions real-time surveillance of online KI targets using information obtained from internet service providers and internet café access control. The report also lists two categories of online KI targets: targets in category

A, likely those deemed serious threats, are subject to long-term surveillance and all kinds of spying techniques; targets in category B, likely individuals considered lesser threats, are to be monitored with "necessary investigative and control measures" so that the police remain aware of their activities.[113] Although the language is vague and allows for police discretion, in practice, this probably means that targets in category B are subject to less invasive measures.

CONTROLLING BATTLEFIELD POSITIONS—whether online or in brick-and-mortar locations such as businesses, universities, and monasteries—is a powerful framework for neutralizing threats to one-party rule. The Chinese surveillance state has deployed this tactic to great effect thanks to its system of distributed surveillance, which maximizes organizational capabilities while addressing the coercive dilemma. As with other security priorities, the goals of battlefield surveillance and control are set by the party through the rules it promulgates, intended to restrict the freedom of action of regime opponents and facilitate surveillance of their activities. Thereafter, the actual work of controlling battlefield positions is left to "boots on the ground": the surveillance state turns to the party's strengths in organization and mobilization. New specialized bureaucracies—monastery management committees, cyber agencies, and cyber police—form quickly to carry out the central state's agenda using a combination of labor- and technology-intensive methods. In the cyber area, technology has an especially strong role, yet even here, informants and police investigations and intimidation are essential. And the party's mobilization capacities really shine when it comes to controlling traditional battlefields such as religious and educational institutions. The continuous application and refinement of the battlefield-control framework again speaks to the comprehensiveness of the regime's approach, as well as its adaptability in responding to emerging threats to its rule.

CHAPTER 7

Upgrading Surveillance

When the CCP crushed the 1989 pro-democracy movement in Beijing, it was a laggard in science and technology, unable to provide its police with modern instruments of surveillance. Surveillance cameras were nowhere to be seen on the streets of the Chinese capital; in general, the police were poorly equipped when it came to technological means of repression. Indeed, they were poorly equipped in many respects, lacking in facilities and vehicles as well.

An officer who retired in 1989 would almost certainly find their old unit unrecognizable today. Once-decrepit police stations have been replaced by spacious, well-furnished buildings with every amenity. The nerve center of the new unit features a wall of flat television screens displaying live images of major traffic routes, shopping centers, and public squares. Instead of bicycles, police now get around in cars bristling with secure wireless communications equipment. Automatic alerts warn duty officers when surveillance cameras with facial recognition technology spot individuals on authorities' blacklists.

The technological great leap forward of the post-Tiananmen era attests to the CCP's resolve to defend its political monopoly and its capacity to recruit resources for high-priority objectives.[1] Although three decades ago few predicted the rise of China's techno-surveillance state, in retrospect several early factors contributed to it. Rapid economic development unleashed forces that could overwhelm even the party's fearsome labor-intensive surveillance apparatus, necessitating improvements. A newly mobile population with expanded access to information tested aging bureaucratic mechanisms and compelled the party-state to pursue modern methods of tracking people and goods. At the same time, the process of creating an information society worked in favor of the surveillance state. The widespread adoption of information technology, in particular mobile communications devices, enabled police to record digital footprints and monitor communications and movements. Ever-adaptable in responding to threats, the party turned a new techno-social challenge into an opportunity by exploiting the vulnerabilities of an always-online population.

Several additional practical factors aligned with the party's post-1989 priorities. Burgeoning revenues provided the funds necessary to acquire new technologies. Friendly commercial relations with the West allowed China to import critical tools and knowledge with few restrictions.[2] At the same time, competitive homegrown tech companies fostered the indigenous capabilities of the Chinese surveillance state. For government bureaucracies and party officials, the technological upgrading of state surveillance also created lucrative rent-seeking opportunities; the process of awarding contracts to build and operate high-tech surveillance systems is rife with corruption, but bribes help to ensure the continued interest of officials in charge.

Further, it stands to reason that rivalries among the various organs of the coercive apparatus have driven the expansion of the techno-surveillance state. A bureaucracy has strong incentives to lobby for its own surveillance system after a rival has acquired new technologies. In the Chinese case, after the MPS completed its Skynet surveillance project in the early 2010s, the CPLC rolled out Sharp Eyes. Although Sharp Eyes includes an effective extension of Skynet, many of its components are redundant.

However impressive, technology is suitable only for certain surveillance tasks. Video surveillance, sensors, and online tracking can, of course, help police better monitor targets' activities. Yet machines complement, but do not replace, labor-intensive surveillance—some routine but essential tasks can only be performed by human beings. For example, research on the use of AI in policing shows that the technology is helpful in developing criminal profiles and assessing risks, but is hard to imagine that AI can fully automate the complex work of, say, political-legal committees, which depends on judgement and political experience.[3] Furthermore, human informants, by virtue of their direct contact with targets, can gather valuable intelligence beyond the reach of state-operated surveillance technologies. This is especially important where targets know how to thwart techno-surveillance. Human beings also are better-equipped than machines to perform certain critical tasks of preventive repression, such as intimidating high-value targets.

What makes the Chinese security apparatus more formidable than those of other dictatorships is not the party-state's recent aggressive adoption of technology but rather the combination of technology and labor in a system of mass surveillance that can maximize the advantages of both. In other words, when we succumb to the hype surrounding technologies like China's proposed social credit system, we are missing the underlying realities of techno-surveillance: it is effective because of labor-intensive systems based in the hukou and the structure of grid management—themselves features of the baojia, a thousand-year-old project of social control.[4]

The Party's Golden Shield

The initial step the Chinese government took to upgrade surveillance involved modernization of MPS and local police information technology. Effective surveillance of a modern society is impossible without a nationally integrated IT infrastructure, consisting of databases containing vast amounts of information about ordinary people, a dedicated and secure communications network connecting widely distributed police personnel, and software tailored to surveillance tasks. There is nothing especially eye-catching about these technologies; secure connectivity, for instance, sounds very basic in this day and age. But without these foundations, a national system could not be achieved.

The first effort at creating such a system was the China Crime Information Center. Established in 1994, it was modeled on the FBI's National Crime Information Center. But at the time the Chinese police lacked the fundamentals noted above: the tools needed to quickly and securely store and share information.[5] There was little the new effort could accomplish.

The next attempt was the MPS's Public Security Informatization Project, also known as Golden Shield (*jindun*), launched in September 1998. (The hope was for completion by 2006, but development continued well beyond that point.[6]) The goal of Golden Shield was to modernize the IT capabilities of security agencies so that they could store and share information and thereby take advantage of high-tech surveillance. Much of Golden Shield, then, was mundane. Engineers developed a dedicated public security information network connecting the MPS to local agencies so that police could transmit data across the country, between community stations, and to far-off headquarters. Police officers got a secure wireless communications network, and new command centers were installed in local police headquarters.[7]

The Golden Shield design was clearly useful for conventional law enforcement, but some of its components were just as obviously suited to political surveillance.[8] The MPS called for development of twenty-three "category-one" specialized software applications (*yinyong xitong*) using data collected by the police or provided by commercial establishments; the agency has not provided a full list of what these applications are, but we know that they include software used to monitor targeted groups.[9] For instance, the Shanghai Public Security Bureau described a category-one tool as a potent surveillance resource. In 2003, the PSB revealed that this system contains data on full-time and temporary residents, rental housing, hotels, and "information on individuals under surveillance classified as 'subjects or targets of work,' including descriptions of cases and incidents."[10]

Known surveillance applications under the aegis of Golden Shield include Domestic Security Intelligence and Information Management, Hotel Law and Public Order Management Information, Public Information Network Security and Surveillance Alarm and Handling, Information on Drug-related Individuals, National Evil Cult Case Management and Analytics, and Management of Foreign Nationals or Individuals Outside Borders.[11] Civil Affairs Information, another special application, focuses on nongovernmental organizations and includes registration information and regulatory-compliance records of municipal and district social organizations.[12]

Alongside the category-one systems mandated by the MPS are tools developed by local public security agencies using Golden Shield funding. These applications are tailored to the specific needs of the agencies responsible. In 2009, the provincial Public Security Department in Guangdong completed its Comprehensive Intelligence Platform Subsystem for the Management of Key Subjects.[13] In 2004, the PSB of Yangzhou, Jiangsu, used Golden Shield funding to develop an application for cross-referencing information. The PSB claims that, by cross-referencing among databases of full-time residents, temporary residents, registered hotel guests, renters, automobile drivers, traffic violators, individuals under police investigation, drug dealers and users, and individuals subject to police surveillance and control, it has advanced its capabilities to surveil and control the online activities of key individuals.[14]

The Golden Shield component specifically designed for cyber surveillance is the Public Information Network Security Surveillance and Control System, popularly known as the Great Firewall. According to a senior MPS official, the purpose of this system is "to ensure the secure operation of public networks, crack down on cybercrimes, and surveil and control all types of harmful online information."[15] From the outset, the CCP decided to give the MPS primary responsibility for online surveillance, and technological capabilities to carry out this mission were installed inside police facilities.[16]

The effectiveness of Golden Shield critically depended on the collection and entry of vast amounts of information into assorted databases. In the parlance of policing in China, these tasks are called "basic work" (*jichu gongzuo*), a term that belies just how challenging the job is: China's police were mobilized in large numbers to make Golden Shield a reality. For example, in 2006, Jiangsu provincial police entered more than 330,000 pieces of information *each day* into Golden Shield databases.[17] Implementation of Golden Shield underscores once again the importance of a preexisting organizational infrastructure in the buildout of a high-tech surveillance state.

Skynet

In 2005, the MPS launched an ambitious and costly video-surveillance project: the City Alert and Surveillance Technological System, or Skynet.[18] Its major features and capabilities can be gleaned from a June 2011 MPS document discussing expansion of the system.[19]

Skynet was conceived as an integrated network using "smart" technology, an apparent reference to the application of big data analytics. According to the MPS, Skynet was designed to be a digital platform on which video information collected by different police departments could be shared. This brought together cameras, other sensors, fiber-optic data links, bespoke software, data servers, and standardized databases enabling sharing. The video cameras would provide wide coverage on their own, but the system would also be able to handle incoming video from cameras on outside networks, allowing integration with surveillance systems operated by institutions apart from the police.[20] This is crucial: many locations escape the view of police cameras, but Skynet would allow police to watch surveillance footage from government agencies, state-owned enterprises, businesses, universities, and other institutions in real time. In this way, police could extend their coverage without installing new equipment.

Skynet mainly relies on cameras to cover key "law and public order" areas. The system also includes sensors and automatic alarms installed at "smart checkpoints" located at border entries and major highway and waterway intersections. The information generated by smart checkpoints can be integrated into a common platform and shared widely. In its Skynet document, the MPS also requires large-scale adoption of "smart, high-definition image-capturing devices," RFID equipment, mobile-phone tracking, and collection of unspecified "trusted information"—most likely personal information stored in RFID-enabled national ID cards.

The MPS document indicates that Skynet networks operated by provincial police departments and city, county, and district PSBs were to be connected to each other, and superior police agencies would have the ability to monitor these subordinate networks. Surveillance platforms at various levels would gradually interface with other information platforms maintained by the police, such as those tracking police emergencies, fire alarms, and traffic accidents. Once the 2011 orders were carried out, police monitoring centers at the city, district, and neighborhood levels would have real-time surveillance capacity using video cameras and sensors.[21]

Initially, the principal function of Skynet was to provide the police with technological capabilities to conduct real-time visual surveillance of streets, highways, and other public venues and to store images. As more advanced technologies have become available, Skynet has acquired additional capabilities. One example is the smart checkpoints—invisible electronic checkpoints equipped with cameras, license plate readers, Wi-Fi sniffers (to collect mobile phone information), and facial recognition technology. These checkpoints enable police to track the movements of vehicles and individuals in real time.[22] A police officer enters into the system the digitized identifying information of an individual, or the license plate of a vehicle, and the individual or vehicle will be quickly located. The way it works is that, when people and vehicles pass a smart checkpoint, Skynet collects license plate, facial recognition, and mobile phone information. This can then be compared with the information stored in police databases to ascertain whether the individual or vehicle is under surveillance. If the individual or vehicle is a target, Skynet automatically alerts police to the target's location.

Local police acquired this capability by the mid-2010s, if not earlier. In 2013, public security agencies in twenty-one of the thirty-one provinces had set up provincial-level video information-sharing platforms. Among 460 city-level PSBs, 332 had video-sharing platforms connected to more than 600,000 networked cameras.[23] The PSB of Huangping County, Guizhou, reports that, in 2017, it used facial recognition systems and electronic surveillance to monitor more than a thousand people of interest, such as repeat petitioners, criminal suspects, fugitives, drug users, ex-convicts, and individuals released from detention. Skynet apparently generated thousands of accurate matches.[24]

Skynet is a high-tech system, but utilizing it is a labor-intensive process: police must continuously watch and analyze the video feeds. To this end, Wuhan's PSB formed a special video unit in 2012. Its ranks quickly swelled to 800 officers and 1,900 civilian monitors.[25] In Weng'an, Guizhou, the county PSB in 2015 set up a "video-image investigation unit" that monitored video twenty-four hours a day. This unit assisted the DSP unit in performing video inspections and in-person interrogations of individuals considered threats.[26] The police chief of Bao'an District, Shenzhen, disclosed that, by the end of the 2000s, his PSB had installed 174 surveillance centers (*jiankong zhongxin*) staffed by 764 full-time police officers and police assistants providing round-the-clock coverage.[27]

The Four Phases of Skynet

The construction of Skynet has been spread out over time, as the system accrues new capabilities. We can gain a better understanding of the various phases of Skynet by looking at the buildout in Liuyang. A county-level city within Changsha, the capital of Hunan Province, Liuyang had a population of 1.5 million as of 2019.[28] Construction of Skynet in Liuyang began in 2011. The first phase likely took two years to complete. During this phase, a contractor, China Telecom's Changsha subsidiary, installed 1,807 high-definition cameras covering 1,274 locations including "key urban sections, traffic nodes, high-crime back alleys, and public transportation venues," mostly in the city center. The cost averaged 18,000 yuan per camera, about $2,500 as of this writing. The second phase began in 2014, with the addition of close to a thousand additional high-definition cameras, extending the coverage of Skynet from the city center to outlying towns.[29] During the third phase, in 2015, police added another 800-plus high-definition cameras and expanded coverage to the main transportation routes and "critical areas" of outlying townships. Another upgrade apparently occurred in 2020, although its details are unknown.[30]

For clues about this phase-four upgrade, we may look to the municipality of Changsha. After spending more than 1 billion yuan since 2011 on the first three phases of Skynet, Changsha's PSB claimed that, by 2020, it had installed more than 70,000 cameras and had connected its surveillance platforms to an additional 85,000 cameras operated by other entities. There were thus 1.5 cameras accessible to police for every 100 residents.[31] In the fourth phase, the Changsha PSB installed more cameras and replaced older cameras with new ultra-high-definition models. This phase also saw the integration of cloud computing, big data, facial recognition, and other visual analytics technologies.[32] It is probable that Liuyang has followed a similar trajectory.

Disclosures from various localities suggest that the pace of Skynet's construction was uneven across jurisdictions; some cities proceeded faster and adopted more advanced technologies earlier than Changsha did. Guangdong appears to have led the buildout of Skynet during the early phases. Between 2005 and 2008, the provincial government invested 12.5 billion yuan in a system that included 920,000 cameras (roughly one camera for every 100 people). By late 2008, provincial police could monitor major highways and maintain a large number of electronic checkpoints.[33] Nanning, the capital of Guangxi Province, completed the first phase of its Skynet in 2011. In 2013 it began upgrading by deploying cloud storage of video footage and by integrating with other databases, such as the Police Geographic Information System and databases storing population data and driver information.[34] In the Chenghua District of Chengdu, Skynet upgrades began in 2014 with expanded coverage and the installation of ultra-high-definition replacement cameras. Facial recognition technology and smart sensors came in 2017, and more cameras were upgraded in 2019.[35]

Financing and Sustainability

Though Skynet is run by the MPS, available official documents contain no references to appropriations for it by the central government. Instead, the costs of building Skynet were apparently borne by local governments. The expenses of Skynet fall into four categories: a network of fiber-optic cables connecting front-end equipment to monitoring centers in the PSBs; hardware such as cameras, sensors, servers, computers, and displays; routine maintenance of the network and hardware; and system upgrades.

Local governments came up with various payment schemes. Some jurisdictions paid upfront. More cash-strapped localities used public-private partnerships, relying on firms to make the initial investment. The government would then lease access to the system.[36] For instance, in Ju County, Shandong, Skynet was built by unspecified "enterprises," leased by the county government, and then operated by the PSB.[37]

For all its surveillance capabilities, Skynet was beset with problems from the outset. When Shanghai built its Skynet in the late 2000s, it incurred enormous expenses. Installing cameras (and probably the associated fiber-optic cables) cost an average of 70,000 yuan apiece, or about $10,000 as of this writing. Every monitoring station required two officers, each working a twelve-hour shift, necessitating the hiring of 900 assistant police at a cost of 30,000 yuan per year. Even excluding future upgrading, the cost of maintaining Skynet for ten years in Shanghai, with annual depreciation of 10 percent, is at least 1.2 billion yuan.[38] Although a wealthy city like Shanghai may be able to afford such costs, less well-off jurisdictions cannot. As one police officer pointed out, the financing model of "telecom builds, government leases, public security uses" ensures that local governments are saddled with a new expense in perpetuity, even as they obtain no new revenue sources to pay for it.[39] Like many high-tech systems, Skynet might have been easier to build than it is to maintain. The example of Wuhan is sobering.

The contractor hired to maintain Skynet in Wuhan had a team of 2,000 workers performing routine upkeep in the mid-2010s.[40]

Alongside financial costs, Skynet has faced significant technological challenges. Interfacing Skynet with other surveillance systems, such as that run by the traffic police, has been a thorny process. In addition, even as Skynet's purpose is to enable sharing of video with and between police, the various systems comprising it were not built to be compatible, because they were not designed to any one standard. Some databases and network protocols were standardized, but others were designed at the local level, and the hardware was assembled by different jurisdictions using different contractors and equipment, creating headaches when it came to integration.[41] Other serious problems have included a lack of well-trained staff to operate the surveillance systems, insufficient coverage, poor maintenance, duplication of efforts and waste of resources, and rapid obsolescence of equipment.[42]

The Party Needs Sharp Eyes

The Sharp Eyes (*xueliang*) project, launched in May 2015, augments Skynet but in many ways is a separate, if not duplicative, surveillance program led and coordinated by local political-legal committees. As evidenced by the official documents analyzed below, Sharp Eyes is best understood as consisting of four distinct components, all of which aim to improve and expand the state's surveillance capabilities. The first component of Sharp Eyes is effectively an upgrade and expansion of Skynet. It is built, operated, and maintained directly by the police. The second component is the installation of video surveillance systems by all state entities, to which the police have access even though these systems are not directly part of Skynet. The third component consists of video surveillance systems operated by nonstate entities; these entities are required both to build monitoring capacity and make the resulting footage accessible to police. The fourth component is an expansion of video surveillance to rural areas, under the authority of local outfits affiliated with the party's political-legal committees.

The political impetus for adding another costly high-tech surveillance program came directly from top leadership. In April 2015, the General Offices of the CCP Central Committee and the State Council issued a joint "Opinion on Strengthening the Construction of Systems of Law and Public Order, Control, and Prevention in Society." The opinion stresses the development and sharing of information resources through use of big data, cloud computing, smart sensors, and other technologies and calls for accelerated construction of video surveillance systems in public spaces, with priority given to improved coverage and video quality in previously neglected rural areas.[43] While the extension of video capabilities to rural areas made sense for a party seeking total control, the gaze of the resulting Sharp Eyes system also fell on urban areas already covered by Skynet.

Why would the party approve an expensive project that replicates so much of Skynet? One possible answer is that the CPLC wanted something like Skynet under its own control. Not only would this foster the surveillance capabilities of the CPLC and its local affiliates, but it would also justify increased budgets and personnel. The CPLC itself primed the pump, making large grants for 148 "demonstration projects" between 2016 and 2018.[44]

The document formalizing Sharp Eyes, issued by the MPS and other ministries in May 2015, contains several notable provisions. First, the Office of Comprehensive Social Management—that is, the CPLC, under one of its several guises per the regime of "two organizational titles with the same staff"—would lead and coordinate the buildout of Sharp Eyes. Second, Sharp Eyes was to be completed by 2020, meaning that by this point a fully networked system of high-definition surveillance cameras would extend across all critical public areas. Third, Sharp Eyes would take advantage of the video-sharing platform maintained by the police; this platform presumably is Skynet, although the document does not specify. In order to meet this demand, government departments would be required to upgrade their video-surveillance systems to make them interoperable with the police platform. Fourth, the police would provide guidance for and supervision of the video-surveillance buildout and would set the network standards for Sharp Eyes. Finally, local governments were to include in their budgets the expenses of building, connecting, and maintaining video surveillance systems in key public areas.[45] The CPLC may have provided seed capital, but most Sharp Eyes costs were borne at the local level.

In January 2018, the CCP Party Center and the State Council called for the extension of Sharp Eyes to the countryside, although the nationwide buildout probably had begun in the second half of the previous year.[46] Technologically, Sharp Eyes upgraded the police-run Skynet with the latest equipment, such as drones, Wi-Fi sniffers, facial recognition, facial-expression recognition, vehicle identification, and mobile-phone tracking.[47]

Video surveillance under Sharp Eyes is broken down into three categories. Monitoring of category-one locations is the responsibility of police, who operate and maintain the relevant systems, which are directly integrated into Skynet: in other words, a key part of Sharp Eyes is an extension of Skynet. Systems covering category-two spots are built and operated by nonpolice government departments or state-affiliated agencies. Such systems are to be connected to the "public safety video image sharing platform" of the police, but not directly to Skynet. Video surveillance systems monitoring category-three locations are built by enterprises, non-profit institutions, shops, and residential communities and are connected to non-Skynet police video platforms. Operation of category-two and -three monitoring is under the purview of so-called comprehensive social management centers (in reality, part of local political-legal committees) rather than PSBs. In rural areas, Sharp Eyes operations are further divided among the county, township, and village levels.[48]

Nongovernment entities like residential communities are said to benefit from Sharp Eyes and so must pay for it themselves: the government tells them to build surveillance systems and connect them to official platforms, but the entities pay for construction, operation, and maintenance on their own. In the case of residential communities, monitoring centers are housed in the offices of the relevant management companies; presumably, building attendants maintain watch.[49] Because police must be able to access the systems monitoring category-two and -three locations, police also must approve the construction of these systems and certify their effectiveness.[50]

The technological architecture of Sharp Eyes is much like that of Skynet. Sharp Eyes has four core components: front-end surveillance equipment (cameras and sensors such as Wi-Fi sniffers and automatic RFID readers); fiber-optic networks for data transmission; software performing a variety of functions, such as data and image analytics and video sharing; and command centers housing data servers.[51] In some areas, Sharp Eyes platforms deploy software specifically designed to monitor key individuals.[52] Cameras used at category-two and -three areas are not connected to facial recognition systems and therefore probably are not high-definition cameras.

Reports from the city of Xi'an indicate at least some of the sorts of locations that fall into category one. These include major traffic routes, tunnels, key bridges, exits and toll stations on select highways, main entries and exits at shopping centers, high-crime areas, public squares, other areas used for large gatherings and religious activities, monuments, the vicinities of government buildings, hospitals, primary and middle schools, telecommunications facilities, airports, railway stations, seaports, entities critical to national security, and entities and locations designated as likely terrorist targets. Category-two and -three areas are considered less important. They include kindergartens, museums, hotels, entertainment establishments, and the interiors of shopping malls.[53]

At this point, Sharp Eyes provides exceptionally broad coverage. In 2018 in Xiamen, a city of slightly more than 4 million people, category-one areas were covered by 30,509 channels—video links, each connected to multiple cameras. Category-two areas were covered by 14,464 channels and category-three areas by 29,266 channels.[54] Other localities report similarly large numbers of channels, indicating that a huge quantity of equipment is deployed, incurring substantial maintenance costs.[55]

Challenges of Operating Sharp Eyes

The parts of Sharp Eyes that have been integrated into Skynet are well financed and built to stringent technological specifications: in all likelihood, they provide impressive capabilities, advancing beyond what would be possible with Skynet alone. The other components of Sharp Eyes—built and maintained not by the

MPS and local police but by other central and local government departments as well as state and private entities—probably are of mixed quality and offer lesser capabilities.

A significant challenge inherent in a complex and sprawling surveillance system such as Sharp Eyes is its unforgiving technological requirements. Minor flaws, such as poor connections, substandard maintenance, and software bugs, can greatly reduce its effectiveness. While Skynet operates primarily in urban areas where budgets tend to be larger and technical support is readily available, Sharp Eyes operates mainly in the countryside, where funds are scarce and able support is hard to find. An article on Sharp Eyes, coauthored by a police officer and an engineer, discloses that some of the system's surveillance cameras cannot focus automatically or precisely capture colors and shapes of objects. Dust, bad lighting, leaves, and spider webs frequently obscure the cameras. Cameras perform poorly at night, and the quality of installation is sometimes shoddy.[56]

Perhaps the greatest challenge facing Sharp Eyes is local governments' lack of fiscal resources, especially in poorer rural areas. Each component of Sharp Eyes—cameras, dedicated fiber-optic cables, monitoring centers—is costly to assemble and maintain. Most rural governments also lack skilled personnel to operate sophisticated surveillance equipment. Poor infrastructure, including an unreliable power supply, is a source of disruptions. Low population density also means that Sharp Eyes costs more per capita in rural areas than in cities.[57] And recall that Sharp Eyes was built on central government orders but without a unified plan for local governments to follow. A study of Yancheng, Jiangsu, for example, finds that officials there used their discretion wastefully, a common problem in an enormous country where top officials issue edicts but cannot easily control how they are carried out.[58]

The Social Credit System

The latest major innovation in China's ecosystem of preventive repression is the proposed national social credit system (*shehui xinyong tixi*). Social credit has gained enormous attention in Western media, and for understandable reasons: the data-hungry system would assign every Chinese citizen a credit score based on evidence of what the state considers prosocial and antisocial behavior and on perceived political loyalty. Using the credit number to dole out rewards and punishments, the government would have new tools to promote the total obedience of the public.

A major question is whether the hype matches reality. And the fact is that we don't really know yet because social credit is a work in progress. The push for a national system gathered momentum after the rise of Xi Jinping in November 2012.[59] By the following summer, the State Council released a planning document announcing that the social credit system would be built during the 2014–2020 period.[60] This did not happen, evidently, because in July 2019, the State Council designated a lead coordinator for system development. But several pilot programs were implemented during the five-year period, and the council endorsed many of the measures that were tested.[61] From official documents and media reports, we can surmise that additional pilots are underway.[62] So the project is not complete, but progress has been and continues to be made. In December 2020, the State Council issued guidance on criteria and procedures for determining creditworthiness.[63] In

March 2022, the party added requirements in hopes that the social credit system will serve new development objectives, such as a growth model centered on consumer demand for domestically produced goods, making China less dependent on access to markets in potentially hostile countries and less vulnerable to possible economic sanctions from abroad. However, the party did not elaborate how exactly the social credit system could help advance these objectives.[64]

Important details about social credit remain unknown. Existing scholarly literature focuses on design, implementation, public perception, and possible applications of the social credit system. But how does the system actually operate? Much speculation concerns the use of social credit for political spying, but there is little solid evidence to go on.[65] There also appears to be a sense that, because social credit is a high-tech initiative, it is especially dangerous and even infallible. But what we do know suggests that, like Sharp Eyes and other surveillance initiatives, social credit as envisioned is subject to diverse local government practices, which will likely create serious complications in building an integrated and reliable national initiative. Due to path dependence, initial flaws in the design of the system could permanently hamper its effectiveness as a surveillance tool.

The system that has been described by authorities would cast a wide net, collecting and processing information not only about ordinary people but also government agencies, officials, businesses, and nongovernmental organizations. The proposed buildout of the social credit system consists of constructing information databases for various regions and economic sectors, the establishment of systems for collecting the relevant information, and the promotion of information exchange across sectors and regions. In keeping with

Leninist organizational imperatives, the state will not be collecting social credit information alone: commercial and other entities will be required to join in gathering, processing, and storing the relevant data.[66]

The social credit system, if it is realized as intended, will be another instance of distributed surveillance. Building it and operating it will call upon the attention and energies of countless people spread across the landscape of Chinese society—not only or even primarily the capacities of state security bureaucracies. The foundation of this data-processing initiative will be the information provided by grid attendants, university staff and students, local party and government officials, business operators, residential management staff, medical personnel, schoolteachers, urban facilities managers, and everyday people going about their lives in the most unassuming manner. Formal and informal informants of all kinds, monitoring their neighbors, colleagues, friends, and relatives, will furnish the data that make a coercive national social credit system possible.

Implementation and Challenges

As of the end of 2021, sixty-two localities had been selected for social credit pilot projects. The first type of pilot, known as a regional comprehensive pilot, is supposed to evaluate the performance of a system within one jurisdiction. The second type, a regional cooperation pilot, tests methods for collaboration and exchange of information as well as coordination of reward and penalty programs across jurisdictions. The third type of pilot, the trial credit-reporting system, tests a full-blown social credit system in the "key sectors" listed in the State Council's 2014 planning document.

One element of the social credit system under construction has already become a powerful instrument of social control: public shaming. One of the punishments envisioned for the system is public humiliation of low-scoring individuals. In some jurisdictions, this is happening today, as courts seek to enforce their judgments and penalize those failing to pay debts by publicizing the names of individuals and companies labeled "untrustworthy." These individuals, as well as individuals responsible for targeted entities, are also subject to penalties such as denial of access to high-end hotels, restaurants, and apartments and to first-class seats on planes and trains. Designated individuals may be prevented from buying real estate, renovating their houses, taking vacations, purchasing pricey insurance products, or sending their children to expensive private schools.[67]

Much of the progress toward a national-scale social credit system has occurred in the development of a legal framework. It appears that, by the end of 2021, governments in nearly all provinces and large municipalities had created social credit regulations.[68] Provisions in these regulations are loosely defined, reflecting the party's long-standing preference for maximum discretion. For example, the Shanghai municipal government announced in 2020 that it would classify as untrustworthy individuals who concealed COVID infections, travel history to pandemic-stricken areas, close contact with COVID patients or suspected patients, and evasion of mandatory medical isolation. The central government's orders said nothing about classifying such people, but they were reported to Shanghai's public credit information platform, which then cut their social credit scores.[69]

Other jurisdictions that have interpreted social credit rules liberally include Guangdong. Article 32 of its provincial regulation, for example, classifies as untrustworthy anyone engaged in conduct that results in "serious sabotage of media order in cyberspace or the gathering of a crowd to disturb social order." The China Cyber Administration has proposed that all jurisdictions blacklist as untrustworthy any individuals who spread online "rumors" that have an "egregious social impact."[70] Article 23 of Nanjing's social credit regulation reduces the creditworthiness of individuals who "drive under the influence, keep violent or aggressive dogs illegally, disrupt order in healthcare facilities, ride public transportation without paying for tickets, [or] organize direct marketing activities . . . that affect social stability."[71]

The social credit system has unmatched potential as a surveillance tool because of the vast amounts of personal data collected, stored, and analyzed under its aegis. A fully functioning social credit system as envisioned by the central government can theoretically use big data and AI to develop relatively precise profiles of individuals' political leanings and even predict the risks that a given person may pose to the party. However, press reports suggest that, so far, the system has been deployed largely as an instrument of social control through administrative penalties, not as a high-tech tool for political spying. This seems implicit in social credit's frequent abuse by local governments. Many jurisdictions joined Shanghai in wielding the penalty of credit reduction to enforce COVID containment measures.[72] Other questionable uses of the social credit system include penalizing individuals who charge their electric bikes in public areas of their buildings of residence, drivers who run stoplights, and litterers.[73] These may not be laudable activities, but social credit is a permanent record that could lead to potentially serious consequences. The profusion of discrediting actions should itself be frightening, as it speaks to the wide discretion of officials. In some areas, governments have penalized petitioners seeking redress of their grievances with loss of social credit.[74]

The proposed social credit system faces two daunting challenges and may ultimately fail to achieve its potential as a surveillance tool. One is a lack of national standards for actions that would affect creditworthiness, positively or negatively. This allows local governments more latitude to impose social control, but the deleterious impact on a national social credit system is likely far-reaching. Information collected in the absence of clear guidelines will contain an elevated level of noise and thus will be unreliable in ascertaining an individual's true political loyalties. Indeed, the agendas of local officials in adopting social credit rules will likely diverge significantly from that of the central government, thus rendering this system less effective.[75] Another challenge is the integration and processing of vast amounts of credit information across jurisdictions and between governments and nongovernment entities.[76] The expense and challenge will be enormous, although that is not to say they will be beyond the capacity of the party-state.

THE UPGRADING OF THE COERCIVE APPARATUS that began in the late 1990s has significantly improved China's surveillance capabilities. As a result, the Chinese system of distributed surveillance has acquired a new technological dimension. The top leadership has supported techno-surveillance generously, investing the returns from rapid economic development. And the Leninist party-state's capacity for mass mobilization means that these investments need not go to waste: organization enables implementation.

However, some technologically advanced surveillance programs are better designed and more effectively implemented and maintained than others. Golden Shield and Skynet, the two programs run exclusively by the MPS, are especially capable and effective, and appear financially sustainable. Sharp Eyes, by comparison, has serious flaws and will likely prove less sustainable. And the regime will encounter difficulties in realizing the aspiration of social credit scoring. Chinese officials have overcome the technological challenges of assembling and integrating across large-scale databases before, but, no matter how good the technology is, this much-feared system is easily undermined by local authorities' opportunism, which transforms a mechanism of surveillance into an imperious tool of social control that could easily alienate citizens.

The Chinese experience with techno-surveillance yields two important lessons. First, a regime with an established organizational infrastructure and tried-and-true labor-intensive approaches to surveillance is likely to apply new technologies more effectively than a regime lacking in such attributes. The adoption and effective utilization of modern surveillance technologies requires concerted political mobilization and administrative coordination that only well-organized dictatorships are capable of delivering. Second, although modern technologies may replace human labor for some functions, they cannot completely automate surveillance. At least, they cannot yet do so, and a fully automated surveillance system will be a long time in coming, even if relevant AI technologies already exist.

For now at least, a technologically sophisticated surveillance state must retain its labor-intensive organizational structure. If China is the global power closest to the dystopic Orwellian ideal, it is not because it has adopted high-tech tools. It is because it has the human infrastructure needed to make good use of those tools.

Conclusion

This study has revealed the institutional framework, principal components, and tactics of distributed surveillance in China, the key to the party-state's coercive apparatus. The foundations of the country's system of distributed surveillance were laid in the Maoist period, but that system was institutionalized, expanded, and modernized after Tiananmen and the economic boom of the 1990s.

Largely through trial and error and learning by doing, the CCP has developed a comprehensive, flexible, and labor-intensive approach to preventive repression that seeks to take maximal advantage of the organizational capabilities of its Leninist institutions. As a result, the Chinese surveillance state is better coordinated and equipped than any other dictatorship in history. China's surveillance state stands out for its multilayered structure, which allocates surveillance tasks among various security agencies and other entities, and for its adaptability, as shown in the party's success in dominating the new "battlefield positions" of Tibetan Buddhist monasteries, university campuses, and the internet. Based on local data, we are able to estimate some key parameters of the Chinese surveillance state, such as the size of its network of informants and the share of the population subject to state-sponsored mass surveillance programs. As more materials about China's surveillance state come to light, we should be able to gain more detailed knowledge about its organization and operations.

One of the hazards facing scholars of dictatorship is that they often become so engrossed in piecing together the empirical puzzle that they lose sight of the big picture. Regrettably, this may be unavoidable. In the case of investigating the Chinese surveillance state, the most important task is arguably that of uncovering crucial evidence to reveal its architecture and key parameters, such as the scope of surveillance and the size of spy networks. This research necessarily entails lengthy and detailed presentation of empirical data at the expense of a thematic narrative. I hope to provide a sense of the big picture as well, though. So let us consider some of the broad questions that emerge from this study, and maybe some answers as well.

Unique Features of the Chinese Surveillance State

Although media attention in recent years has focused on the high-tech features of China's surveillance state, our study shows that the adoption of high-definition video, facial recognition tools, and online censorship came relatively late: these technologies strengthened the capabilities of an already-formidable surveillance state. In reality, the keys to the Chinese surveillance state's far-reaching effectiveness are not technologically intensive. They are labor and organization intensive. Crucial in enabling the organizational foundations of the surveillance state are China's Leninist institutions.

Perhaps the most striking feature of the Chinese surveillance state is its multilayered structure. Unlike many other dictatorships, which aggravate the security dilemma by relying on a single secret police agency, China has built three lines of defense against political threats and established a relatively clear division of labor among them. The Ministry of State Security guards the regime against external threats but also assists its domestic secret police, the Domestic Security Protection unit of the Ministry of Public Security. DSP units focus primarily on political threats, whereas routine surveillance operations are assigned to police stations, the outer layer of the coercive apparatus.

Then, beyond the core of the surveillance state, is a vast network of informants who provide intelligence and information. This peripheral layer of the surveillance state consists of institutions and organizations directly controlled by the party-state, such as neighborhood committees, state-owned enterprises, government bureaucracies, state-affiliated social organizations (such as official labor unions and religious groups), and universities. Officials and security personnel in these organizations assist the surveillance state by recruiting informants, maintaining routine surveillance, and performing enhanced security operations during sensitive periods.

On the surface, this multilayered system of distributed surveillance appears to be redundant and costly. In some respects, it is. But the redundancy is probably intentional: the CCP seeks maximum regime security, and it is willing to spend whatever resources are necessary to buy "insurance policies."

Another feature that sets China apart from other dictatorships—apart even from fellow Leninist regimes—is a specialized party bureaucracy that supervises and coordinates the activities of the coercive apparatus. Here I am referring to the political-legal committees, under the umbrella of the Central Political-Legal Committee. The CPLC and the local committees enable the CCP's overall approach to surveillance, which can be summed up by the term *juguo tizhi*: the "whole-of-government and whole-of-society method of mobilization." A Leninist regime possesses unrivaled capacity for mobilizing resources. But mobilization without effective supervision and coordination results in waste. The political-legal committees work to ensure that China does not waste the resources it mobilizes. Critically, by enabling distributed surveillance and thereby preventing concentrations of power forming within particular security agencies, political-legal coordination has helped the party address the coercive dilemma.

Unquestionably, China does possess the most advanced surveillance technologies among all dictatorships. The technologies at the disposal of China's coercive apparatus have significantly upgraded its capabilities to track the communications and activities of known and potential threats to the party. Yet the Chinese experience also illustrates the potential limitations of modern surveillance technologies. A regime's organizational capacity is a precondition for the effective deployment of surveillance technology. China has been successful in responding to the information revolution with its Great Firewall and in adopting modern surveillance technologies because the party-state had an existing organization-intensive surveillance structure in place before these technologies were available. All the regime had to do was equip an already-formidable surveillance apparatus with more advanced tools. Placing modern technologies in the hands of a poorly organized surveillance state is sure to produce inferior outcomes.

The Chinese case also shows that technologies may complement, but not substitute for, human labor. They may expand the scope of surveillance and perform certain functions more efficiently than an unaided human can, but a fully automated surveillance state is still the stuff of science fiction. If anything, the Chinese case shows that the adoption of new technologies requires an enlarged and better-trained workforce, not fewer people. Revealingly, the size of the Chinese surveillance state has not shrunk because of its access to advanced technologies. Instead, the number of personnel devoted to security has grown even as more technologies have been added to the suite of coercive tools.

These observations suggest that it would be virtually impossible for non-Leninist dictatorships to acquire the same surveillance capabilities as those that exist in China. Such states may be able to build a core component like a secret police agency, but not the other vital and complementary elements we find in China's surveillance state. Non-Leninist regimes simply do not have the same organizational infrastructure, comparable control of the economy and society, or similar capacity to mobilize resources. Even if non-Leninist dictatorships were to acquire China's modern surveillance technologies, they would not possess the institutional prerequisites to use these technologies effectively.

It is my contention that embedding surveillance capabilities within preexisting Leninist institutions explains why rapid economic development under such a regime may not lead to democratizing political change. Although economic development produces structural changes favorable to democracy, it also generates the resources—in particular, increased revenue and access to technologies—that allow Leninist regimes to adapt and strengthen their surveillance capabilities. A Leninist regime that can gain wealth without sacrificing its key political and economic institutions to reform will benefit from economic modernization: its hold on power will be strengthened, not undermined.

That is, until economic modernization itself falters. The conclusion to be drawn from this observation is that a viable path of transition from a Leninist regime to a democracy starts with political reforms that uproot Leninist institutions, not with economic liberalization and modernization.

Assessing China's Surveillance State

Based on several empirical measures, China's surveillance state during the post-Tiananmen era has largely fulfilled its mission to the party's satisfaction. Without an effective surveillance state, the party could not have prevented the rise of an organized opposition; contained social unrest; suppressed Falun Gong, the largest spiritual group to emerge in China in the post-Mao era; or neutralized the liberalizing trends produced by rapid economic development. But it would be a mistake to give China's surveillance state all the credit for keeping the party in power. The post-Tiananmen economic boom is doubtless a crucial factor contributing to the party's survival. A regime enjoying robust performance legitimacy, such as the CCP since Tiananmen, generally has fewer enemies than a regime without such legitimacy, making the job of the surveillance state much easier.

If the economic boom itself is the key to regime survival, one has to ask whether the Chinese surveillance state is too big, targets too many people, or performs too many unnecessary tasks. Perhaps the party does not need the duplicative elements of Sharp Eyes, or it could get away with fewer people under surveillance in the Key

Individuals program. The network of informants also appears excessively large because, as the research here shows, the number of known or suspected political threats is relatively small. Most importantly, the post-1989 economic boom appears ultimately to have undergirded social stability. The 1990s saw challenges due to increased mobility and information access, yet, in time, greater prosperity looks to be a handmaiden of docility and political apathy. No truly threatening source of social unrest has arisen in the period of growing prosperity.

So, is the surveillance state too large? No—not if the party seeks absolute regime security. Its paranoia will not allow it to tolerate *any* risk of losing power. Absolute regime security requires that the party nip any threat in the bud, necessitating a large surveillance state perpetually on high alert. From the party's perspective, absolute regime security is worth every yuan spent, even when marginal investments yield no returns. The party need only behave with financial discipline if the coffers of the Chinese state can no longer underwrite absolute regime security.

It is also arguably true that the direct costs of political surveillance are affordable. Consider that China's surveillance state, like those in other countries, is structured to perform not only political spying but also conventional law enforcement functions; surveillance programs designed to track political threats are performed by a surveillance infrastructure whose capacities are mostly devoted to traditional law enforcement goals. A stronger criticism might be that, even if political surveillance is affordable, it is excessive in that it victimizes ordinary people while preventing them from seeking redress through legal channels, thereby fostering public resentment and potentially the very political threat that the party seeks to contain.

Additionally, it is worth keeping in mind that one cost of political surveillance is opportunity cost. A pair of eyes watching political threats means a pair of eyes not watching threats to public safety or welfare. In other words, the party's imperative of regime security results in the diversion of policing resources. To pick just two examples, human trafficking and food safety are both major concerns in China; perhaps if legal authorities and informants were devoted to these areas rather than political policing, the Chinese people would be better off.

The Surveillance State and the Future of CCP Rule

The overarching practical question raised by this study is whether a powerful surveillance state will perpetuate CCP rule. Given the known capabilities of the Chinese surveillance state, one may be tempted to conclude that organized opposition cannot possibly emerge, at least on a regime-threatening scale. Such certainty may be unwarranted, however. Thus far, surveillance has been effective in preventing and preempting antiregime collective action. But there are reasons to question the long-term utility of surveillance as the primary instrument of the CCP's survival. Below, I address four such reasons.

First, a surveillance state may perform effectively in a relatively stable environment in which standard operating procedures are followed. But effectiveness tends to deteriorate in a moment of crisis. Dictatorships engulfed in crises, such as popular uprisings, leadership schisms, or sudden economic shocks, experience great difficulties maintaining clear communications with their security forces. Coordinating security operations becomes even more challenging than it typically is, and the incentives of agents change as they become focused on individual calculations. Some may want to stick with the regime, while others may hedge their bets. In a crisis, confusion and fence-sitting will unavoidably degrade the capabilities of the surveillance state and endanger regime security.

Second, a surveillance state in a dictatorship is designed to monitor, intimidate, and control a relatively small share of the population. What if antiregime forces achieve a critical mass? Then the task of coercion becomes much harder. Our study shows that, in China, perhaps 1 percent of the population is under routine surveillance for political or nonpolitical reasons. The apparatus responsible for monitoring them can be overwhelmed. The experience of the former Soviet bloc in 1989 shows that, in the face of politically mobilized masses, even a Stasi or KGB can do little.

Third, because the post-Tiananmen surveillance state was built with enormous investments in manpower and technology, its sustainability cannot be assured. One possibility is that, as the Chinese economy slows due to population aging, the party's reversal of promarket reforms, and economic "decoupling" from the West, the state will have fewer resources with which to continually upgrade and expand surveillance. The same economic stagnation that undermines the surveillance state will likely cause rising social discontent, such that the regime loses its guardian just when it needs it most.

Finally, coercion in general, and surveillance in particular, is one of several tools upon which dictatorships rely for survival, and coercion functions most effectively when the dictatorship's other tools are also in good order. Propaganda, nationalism, some other ideology, or material incentives may motivate security agents and their informants. The unity of the ruling elite may raise the quality of the security forces and head off their politicization. Superior economic performance may generate the fiscal resources needed to modernize the surveillance state and reward its agents. Cooptation of social elites may help isolate opponents of the dictatorship, thus facilitating state surveillance. And any of these mechanisms may fail. It is difficult to imagine a high-performing surveillance state surrounded by an otherwise-atrophying dictatorship. When dictatorships fall, it is usually not because of the incompetence of their secret police; other policies falter, dragging down the regime. As the adage goes, "The stone age did not end because the world ran out of stones"; likewise, dictatorships do not fall because their spies stop spying.

The impressive performance of the Chinese surveillance state under normal conditions is enough to vindicate the CCP's hardline survival strategy in the post-Tiananmen era. But there are sobering realities to keep in mind. The very effectiveness of the party's surveillance state may lead to neglect of greater threats to its hold on power, such as pervasive corruption, socioeconomic inequalities, inefficient state-capitalism, and the exclusion of the growing middle class from governance.

Then too, the greatest threat to the CCP political monopoly may be its own coercive power. Should the party continue its present course of neo-Stalinist rule under Xi Jinping, it may find that it has no alternative but to depend more and more on coercion and surveillance to remain in power. This is always a bad sign for a dictatorship; regime strength is reflected in a coercive apparatus that gets little use. Over the years, China's reliance on coercion has ebbed and flowed. These days, we are seeing more and more. The CCP regime would be well advised that the heaviest hand is also the weakest.

APPENDIX: INFORMANTS AND SURVEILLANCE TARGETS

NOTES

ACKNOWLEDGMENTS

INDEX

Appendix: Informants and Surveillance Targets

TABLE I

Informants as Share of Population

Jurisdiction	Year	Informants per hundred population
Pingdinshan, Henan	2019	0.38
Xinmi, Henan	2012	0.06
Tongzhou District, Nantong, Jiangsu	2004	0.76
Baoying County, Jiangsu	2000	1.41
Qinhuai District, Nanjing, Jiangsu	2015	1.39
Lianyun District, Lianyungang, Jiangsu	2011–2013	0.42[a]
Putuo District, Shanghai	2016	0.46
Yueyanglou District, Yueyang, Hunan	2016	0.64
Wu Township, Tongxiang, Zhejiang	2015	4.68
Longfang Township, Huangling County, Yan'an, Shaanxi	2014	1.03
Gaoyao, Guangdong	2014	1.40
Yuncheng District, Yunfu, Guangdong	2013	0.40
Yun'an District, Yunfu, Guangdong	2012	1.32
Nanhuaxi Street, Haizhu District, Guangzhou	2012	1.86
Daowai District, Harbin, Heilongjiang	2014	0.07
Shunyi District, Beijing	2013	2.16
Haidian District, Beijing	2014	0.47
Xicheng District, Beijing	2014	1.32
Tongjiang County, Sichuan	2014	0.15
Diqing Prefecture, Yunnan	2014	0.71
Daguan County, Zhaotong, Yunnan	2010	0.25

Nanzhang County, Hubei	2014	0.73
Dahe Township, Tongzi County, Guizhou	2013	0.43
Zhengding County, Hebei	2012	1.80
Lianmuqin Township, Shanshan County, Xinjiang	2010	1.60
Wushi County, Xinjiang	2012	6.89
Qu County, Sichuan	2012	0.23
Tongchuan District, Dazhou, Sichuan	2009	0.88
Chengdu, Sichuan	2009	0.31
Wusheng County, Sichuan	2012–2013	1.10[a]
Average		1.13
Median		0.73

a Multiyear average

TABLE 2

Output of Informants in Various Jurisdictions

Jurisdiction	Year	Number of informants	Pieces of information and intelligence reported	Pieces of information and intelligence per informant
Baoying County, Jiangsu	2000	12,946	3,523	0.27
Tongzhou District, Nantong, Jiangsu	2004	9,643	7,023	0.73
Lianyun District, Lianyungang, Jiangsu	2014	1,100	3,500	3.18
Qinhuai District, Nanjing, Jiangsu	2015	9,698	6,240	0.64
Pingdinshan, Henan	2019	20,848	2,458	0.12
Jiande, Zhejiang[a]	2011	1,717	1,256	0.73
Wu Township, Tongxiang, Zhejiang	2015	2,681	2,058	0.77
Yun'an District, Yunfu, Guangdong	2012	3,651	1,227	0.34
Putuo District, Shanghai	2016	4,153	612	0.15
Putuo District, Shanghai[b]	2019	277	33	0.12
Tongjiang County, Sichuan	2014	1,050	233	0.22
Gaoyao, Guangdong	2014	10,832	1,174	0.11
Shunyi District, Beijing	2013	13,000	220,000	16.92
Average excluding Shunyi and Lianyun Districts				0.38
Average including Shunyi and Lianyun Districts				1.87

a Report indicates that all informants were owners of small shops.

b Report indicates that all informants were delivery personnel.

TABLE 3
Types of Intelligence Collected by Domestic Security Protection Units in Various Jurisdictions

Jurisdiction	Year	Enemy intelligence	Political intelligence	Social intelligence
Jinzhou, Liaoning	1988	13	28	1,081
Luonan County, Shaanxi	1998	9	63	90
Songyuan, Jilin	1999	60	10	333
Zhashui County, Shaanxi	1998	7	22	96
	1999	16	23	96
	2000	4	6	116
	2001	1	18	28
	2002	1	9	84
	2003	1	22	95
Heshuo County, Xinjiang	2011	56	13	42
Miyi County, Sichuan	2008	6	158	354
	2009	1	117	379
	2010	1	164	354
	2011	2	304	476
	2014	0	205	414
	2015	0	298	193
Dongpo District, Meishan, Sichuan	1998	4	93	425
	2000	2	41	451
	2001	8	73	468
	2002	5	58	336
	2004	11	74	287
	2006	9	52	132
Jiuzhaigou County, Sichuan	2002	4	9	14
Nanxi County, Sichuan	2009	6	14	130
Hezhang County, Guizhou	2009	4	38	60

Tiandong County, Guangxi	2003	3	6	34
	2004	6	4	21
	2005	16	4	45
	2006	3	7	59
	2010	1	5	84
	2011	5	7	76
Total		265	1,945	6,853
Share of total (%)		3	21	76

Note: Two classification systems are used. In some jurisdictions, intelligence as classified as enemy, political, and social. In other jurisdictions, intelligence is classified only as A, B, or C. A likely corresponds to enemy intelligence, B to political intelligence, and C to social intelligence. This table presumes as much.

TABLE 4

Quality of Intelligence, by Jurisdiction

Jurisdiction	Year	Pieces of intelligence collected	Pieces used and reported[a]
Yulin, Shaanxi	2006	630	188
Baoying County, Jiangsu	2000	3,523	175
Dazhu County, Sichuan	2009	853	39
Daocheng County, Sichuan	2008	85	42
Beichuan Qiang Autonomous County, Sichuan	2015	420	200
Zhuanglang County, Gansu	2018	242	83
	2017	456	114
Wuhan, Hubei	2001	993	151
	2003	3,132	621
Hulan District, Harbin, Heilongjiang	2014	213	107
Sartu District, Daqing, Heilongjiang[b]	1986–2005	696	190
Chaoyang, Liaoning	2017	240	175
Yiyang County, Jiangxi	2013	533	483
Chongyi County, Jiangxi	2014	613	363
Chengwu County, Shandong	1995	36	5
Fuyang, Anhui	2005	279	87
Ningxiang, Hunan	2014	1,160	83
Hezhang County, Guizhou	2009	185	102
Zhengzhou Railway Bureau, Henan	2015	2,972	196
Linfen, Shanxi	2017	21,280[b]	5,880
Panshi, Jilin	2001	43	26
Share of collected intelligence utilized and reported (%)			24.1

a Reporting to same-level or superior authorities (本级和上级) only.

b This district reported only intelligence pertaining to political security and social stability. Intelligence reported by other jurisdictions is not specified and presumably covers a wide range of subjects.

TABLE 5
Key Populations as Share of Total Population in Selected Jurisdictions (1980s)

Jurisdiction	Year	Share of population (%)
Xinning County, Hunan	1981	0.19
	1985	0.29
Shaanxi Province	1984	0.65
Qinshui County, Shaanxi	1987	0.26
Heilongjiang Province	1981–1982	0.14
	1983	0.31
	1984	0.59
	1985	0.64
Wangkui County, Heilongjiang	1987	0.38
Jianhua District, Qiqihar, Heilongjiang	1985	0.9
Xiangtan, Hunan	1983	0.28
	1985	0.26
Changchun, Jilin	1986	0.06
Yanji, Jilin	1986	0.95
Dalian, Liaoning	1981	0.07
Manasi County, Xinjiang	1986	0.13
Hangzhou, Zhejiang	1983	0.58
	1987	0.57
Xiangshan County, Zhejiang	1985–1989	0.49[a]
Daishan County, Zhejiang	1981–1982	0.08
	1983–1989	0.39[a]
Fuyang County, Zhejiang	1983–1986	0.52[a]
Jinhua, Zhejiang	1983–1986	0.34[a]
Yongkang, Zhejiang	1986–1989	0.37[a]
Jiande, Zhejiang	1981–1982	0.02[a]
	1983–1989	0.35[a]

Cixi, Zhejiang	1982	0.02
	1983, 1985, 1987	0.41[a]
Jinyun County, Zhejiang	1982	0.05
	1983–1989	0.37[a]
Zhoushan, Zhejiang	1980–1982	0.047[a]
	1983–1989	0.49[a]
Chongqing, Sichuan	1984	0.40
Average		0.35
Median		0.35

a Multiyear average

TABLE 6

Key Populations as Share of Total Population in Selected Jurisdictions (1990s)

Jurisdiction	Year	Share of population (%)
Beijing	1998	0.53
Tongling, Anhui	1990	0.49
Nan'an District, Chongqing	1993	0.82
Qian'an County, Hebei	1997	0.56
Xinye County, Henan	1998	0.51
Inner Mongolia	1999	0.33
Pingjiang County, Hunan	1996	0.18
Leiyang County, Hunan	1992	0.21
	1994	0.24
Xinning County, Hunan	1992	0.30
	1995	0.44
Huitong County, Hunan	1992	0.38
Tongcheng County, Hubei	1994–1995	0.39
Huangshi, Hubei	1994	0.32
Zaozhuang, Shandong	1996	0.28
Pinglu County, Shanxi	1995	0.98
Qinshui County, Shanxi	1992	0.52
Licheng County, Shanxi	1996	0.41
Anyi County, Jiangxi	1995–1997	0.37[a]
Ningdu County, Jiangxi	1991–1994	0.32[a]
Dalian, Liaoning	1990	0.48
Changtu County, Liaoning	1996	0.44
Fujin, Heilongjiang	1994	0.70
	1997	1.04
Jianhua District, Qiqihar, Heilongjiang	1995	1.04
Sartu District, Daqing, Heilongjiang	1999	0.3
Zhashui County, Shaanxi	1998	0.34
Shangnan County, Shaanxi	1991	0.57

Nanzheng District, Hanzhong, Shaanxi	1997	0.32
Longyan County, Fujian	1996	0.33
Yong'an, Fujian	1993	0.63
Hongkou District, Shanghai	1994	0.66
Chongming County, Shanghai	1992	0.32
	1994–1995	0.30[a]
	1997–1999	0.25[a]
Manasi County, Xinjiang	1995	0.15
Tiandong County, Guangxi	1994–1997	0.54[a]
Laibin, Guangxi	1992	0.44
Xiangzhou County, Guangxi	1992	0.89
Xuzhou, Jiangsu	1997	1.1
Yixing, Jiangsu	1991	0.22
Huaiyin, Jiangsu	1996	0.95
Shuangliu County, Sichuan	1994	0.22
Dazhu County, Sichuan	1993	0.34
	1995	0.28
Chenghua District, Chengdu, Sichuan	1991–1998	0.24[a]
Linjiang, Jilin	1995	0.71
Baoshan, Yunnan	1995	0.21
Hangzhou, Zhejiang	1995	0.40
Cixi, Zhejiang	1994	0.50
Yuyao cty, Zhejiang	1990	0.36
Jinyun, Zhejiang	1990–1991	0.51[a]
Average		0.47
Median		0.40

a Multiyear average

TABLE 7

Key Populations as Share of Total Population in Selected Jurisdictions (2000s)

Jurisdiction	Year	Share of population (%)
Cangnan County, Zhejiang	2001	0.27
Beihai, Guangxi	2000	0.10
Anding County, Hainan	2008	0.10
Zhangye, Gansu	2004	0.40
Xinning County, Hunan	2004	0.27
Qinshui County, Shanxi	2003	0.26
Lingbao, Henan	2000	0.09
Longyan, Fujian	2002	0.33
Dehua County, Fujian	2008	0.26
Hanjiang District, Putian, Fujian	2009	0.15
Wuhu County, Anhui	2003	0.19
Qidong, Jiangsu	2000	0.44
Xinpu District, Lianyungang, Jiangsu	2001	0.59
Tongzhou District, Nantong, Jiangsu	2005	0.42
Guannan County, Jiangsu	2000	0.32
Huai'an, Jiangsu	2000	0.53
Changning County, Sichuan	2009	0.14
Xuzhou District, Yibin, Sichuan	2007	0.14
Tongchuan District, Dazhou, Sichuan	2007	0.11
Santai County, Sichuan	2001	0.26
Chengdu, Sichuan	2006	0.21
Chenghua District, Chengdu, Sichuan	2000–2005	0.27[a]
	2008	0.25
Jinjiang District, Chengdu, Sichuan	2007–2009	0.31[a]
Jimo District, Qingdao, Shandong	2009	0.41
Pingyuan County, Shandong	2008	0.12

Wendeng District, Weihai, Shandong	2007	0.17
	2009	0.18
Dongying District, Dongying, Shandong	2005	0.07
Tai'an, Shandong	2001	0.18
Nankang District, Ganzhou, Jiangxi	2003	0.11
Yiyang County, Jiangxi	2006	0.15
Xinjian County, Jiangxi	2002	0.13
Hulan District, Harbin, Heilongjiang	2007–2008	0.16
Daxinganling Prefecture, Heilongjiang	2001	0.12
Yichun, Heilongjiang	2005	0.96
Fujin, Heilongjiang	2002–2003	0.82
Sartu District, Daqing, Heilongjiang	2004	0.24
Jianhua District, Qiqihar, Heilongjiang	2003	0.53
Hinggan League, Inner Mongolia	2001	0.25
Jungar Banner, Inner Mongolia	2005	0.13
Wuhai, Inner Mongolia	2000	0.26
Changji, Xinjiang	2009	0.12
Yunnan Province	2008	0.23
Daguan County, Yunnan	2004–2005	0.28[a]
	2008	0.22
Luxi County, Yunnan	2004	0.21
Weixin County, Yunnan	2001	0.31
	2004	0.48
Luoping County, Yunnan	2001	0.29
Simao District, Pu'er, Yunnan	2002	0.27
Ludian County, Yunnan	2008	0.13
Weinan, Shaanxi	2006	0.25
Zhashui County, Shaanxi	2001–2002	0.22[a]
Xixiang County, Shaanxi	2002	0.43

Nanzheng District, Hanzhong, Shaanxi	2004	0.37
Yinchuan, Ningxia	2005	0.31
	2007	0.35
Longhua District, Shenzhen, Guangdong	2009	0.36
Enping, Guangdong	2006	0.29
Benxi, Liaoning	2001	0.35
	2006	0.31
Chaoyang, Liaoning	2003	0.26
Pingquan County, Hebei	2006	0.07
Liuhe County, Jilin	2009	0.24
Average		0.27
Median		0.26

a Multiyear average

TABLE 8
Key Populations as Share of Total Population in Selected Jurisdictions (2010s)

Jurisdictions	Year	Share of population (%)
Pu'an County, Guizhou	2016–2017	0.51[a]
	2019	0.49
Weng'an County, Guizhou	2013	0.62
	2015	0.90
Qiandongnan Miao and Dong Autonomous Prefecture, Guizhou	2013	0.31
	2017	0.40
Songtao County, Guizhou	2012	0.18
Guiyang, Guizhou	2017	0.99
Nanming District, Guiyang, Guizhou	2018	1.01
Qingrong County, Guizhou	2017	0.61
Tongcheng County, Hubei	2014	0.49
Heihe, Heilongjiang	2013	0.17
	2017	0.19
Yuzhong County, Gansu	2011	0.40
Jinchuan District, Jinchang, Gansu	2010–2011	0.51[a]
Zhengning County, Gansu	2013	0.24
Uqturpan County, Xinjiang	2010–2011	0.24[a]
	2014	0.17
Yiwu County, Xinjiang	2010	0.19
Hami, Xinjiang	2013	0.10
Tacheng, Xinjiang	2014	0.13
Dehua County, Fujian	2010	0.33
Kangding, Sichuan	2019	1.17
Changning County, Sichuan	2011	0.16

Jinjiang District, Chengdu, Sichuan	2011	0.22
	2018	0.11
Youxian District, Mianyang, Sichuan	2015	0.14
Mianzhu, Sichuan	2014	0.23
Baqiao District, Xi'an, Shaanxi	2012	0.12
Urad Rear Banner, Inner Mongolia	2016	0.26
Heping District, Tianjin	2010–2011	0.14[a]
Kenli District, Dongying, Shandong	2014	0.17
Wudi County, Shandong	2014	0.21
Juye County, Shandong	2017	0.18
Baofeng County, Henan	2010	0.43
Yucheng County, Henan	2012	0.27
Wuhu, Anhui	2011–2016	0.42[a]
Pu'er, Yunnan	2014	0.02
Cangnan County, Zhejiang	2011	0.34
Average		0.35
Median		0.24

a Multiyear average

TABLE 9
Ratio of Key Individuals (KI) to Key Populations (KP) in Various Jurisdictions

Jurisdiction	Year	Number of people in KP program	Number of people designated KI	KI as share of KP (%)
Anlong County, Guizhou	2016	1,516	3,815	252
Pu'er, Yunnan	2014	525	877	167
Dehua County, Fujian	2010	1,040	1,373	132
Wuhu County, Anhui	2011	1,138	1,141	100
	2012	1,470	1,491	101
	2013	1,579	1,819	115
	2014	1,584	1,643	104
	2015	1,444	2,034	141
	2016	1,694	2,217	131
Utra Rear banner, Inner Mongolia	2016	155	175	113
Qiandongnan Miao and Dong Autonomous Prefecture, Guizhou	2013	14,539	1,041	7
	2017	19,121	2,008	11
Heihe, Heilongjiang	2013	2,177	2,827	130
Bayan County, Heilongjiang	2010	946	749	79
Weng'an County, Guizhou	2013	3,449	4,501	131
	2015	4,353	5,858	135
Jinglong County, Guizhou	2017	2,095	656	31
Lianshui County, Jiangsu	2002	4,322	4,769	110
Chenghua District, Chengdu	2005	1,359	1,333	98
Heping District, Tianjin	2010	407	2,172	534

Shizhong District, Neijiang, Sichuan	2015	1,459	9,214	632
Average				155
Median				115

Data Sources

Table 1: *平顶山年鉴 2020*, 34, 152; *新密年鉴 2013*, 19, 101; *通州年鉴 2005*, 116, 290; *宝应年鉴 2001*, 26, 88; *秦淮年鉴 2016*, 106, 279; *连云年鉴 2014*, 34, 96; *连云年鉴 2013*, 44, 109; *连云年鉴 2012*, 45, 117; *普陀年鉴 2017*, 1, 48; *岳阳楼区年鉴 2017*, 53, 176; *桐乡年鉴 2016*, 312, 315; *黄陵年鉴 2015*, 367–368; *高要年鉴 2015*, 31, 111; *云浮市云城区年鉴 2014*, 35, 131; *云安年鉴 2013*, 112, 184; *海珠年鉴 2013*, 112, 184; *道外年鉴 2015*, 34, 98; *北京顺义年鉴 2014*, 28, 154; *北京西城年鉴 2015*, 161, 414; *北京海淀年鉴 2015*, 126, 425; *通江年鉴 2015*, 31, 73; *迪庆年鉴 2015*, 233–234; *大关年鉴 2011*, 83, 172; *南漳年鉴 2015*, 46, 108; *桐梓年鉴 2014*, 217–218; *正定年鉴 2013*, 65, 129; *鄯善年鉴 2011*, 275; 2010, 255; *乌什年鉴 2013*, 34, 104; *渠县年鉴 2013*, 73, 148; *通川年鉴 2010*, 62; 2012–13, 52; *成都年鉴 2010*, 2, 86; *武胜年鉴 2014*, 49, 112; *武胜年鉴 2013*, 41, 111.

Table 2: *宝应年鉴 2001*, 88; *通州年鉴 2005*, 116; *连云年鉴 2014*, 96; *秦淮年鉴 2016*, 279; *平顶山年鉴 2020*, 152; *建德年鉴 2012*, 289; *桐乡年鉴 2016*, 312, 315; *云安年鉴 2013*, 184; *普陀年鉴 2017*, 48; *普陀年鉴 2020*, 103; *通江年鉴2015*, 73; *高要年鉴 2015*, 111; *北京顺义年鉴 2014*, 15.

Table 3: *锦州年鉴 1989*, 126; *洛南年鉴 1996–1999*, 289; *松原年鉴 2000*, 103; *瓮安年鉴 2000–2003*, 123; *柞水年鉴 1998–2002*, 123; *柞水年鉴 2004*, 74; *和硕年鉴 2012*, 112; *米易年鉴 2009*, 72; *米易年鉴 2010*, 73; *米易年鉴 2011*, 83; *米易年鉴 2012*, 87; *米易年鉴 2015*, 100; *米易年鉴 2016*, 113; *眉山年鉴 1999*, 169; *眉山市东坡区年鉴 2001*, 135; *眉山市东坡区年鉴 2002*, 113; *眉山市东坡区年鉴 2003*, 99; *眉山市东坡区年鉴 2005*, 131; *眉山市东坡区年鉴 2007*, 117; *九寨沟县年鉴 1999–2005*, 297; *南溪年鉴 2010*, 86; *赫章年鉴 2010*, 290; *田东年鉴 2003–2006*, 178; *田东年鉴 2010–2011*, 152.

Table 4: *榆林年鉴 2007*, 92; *宝应年鉴 2001*, 88; *大竹年鉴 2010*, 132; *稻城年鉴 2003–2008*, 184; *北川羌族自治县年鉴 2016*, 107; *庄浪年鉴 2018*, 156; *庄浪年鉴 2019*, 166; *武汉公安年鉴 2002*, 60; *武汉公安年鉴 2004*, 79; *呼兰年鉴 2015*, 127; *大庆市萨尔图区志 1986–2005*, 393; *朝阳年鉴 2019*, 137; *弋阳年鉴 2014*, 124; *崇义年鉴 2015*, 114; *成武年鉴 1996*, 99; *阜阳市年鉴 2006*, 213; *宁乡年鉴 2015*, 84; *赫章年鉴 2010*, 290; *郑州铁路局年鉴 2016*, 322; *临汾年鉴 2018*, 121; *磐石市志 1991–2003*, 35.

Table 5: "湖南新宁司法篇," http://city.sina.com.cn/city/t/2011-08-17/170221463.html; *中国人口统计年鉴* 1988, 634; *湖南年鉴* 1986, 86; *陕西省志：*公安志, 559–561, *陕西年鉴* 1987, 7; *沁水县志* 1986–2003, 484; *黑龙江省志：* 公安志, 375–377; *望奎县志* 1986–2005, 557; *齐齐哈尔市建华区志* 1996–2005, 473, 633; *莲城金盾* (湖南湘潭市第二印刷厂, 1999), 191; *中国人口统计年鉴*1988, 178; *长春市志:* 公安志, 504; *延吉*

市志, 55, 255; *大连市志：公安志*, 254; *大连统计年鉴* 2017, 94; *玛纳斯县志* 1986–2010, 117, 849; *杭州市人民公安志*, 211–212; *象山县公安志*, 183, 200; *岱山县公安志*, 159; *富阳县公安志*, 239, 260; *金华市公安志*, 195; *永康市公安志*, 88; *建德市公安志*, 194, 261; *慈溪市公安志*, 238, 252–253; *缙云县公安志*, 170, 178; *舟山市公安志*, 211; Ch. 7, *重庆市志：公安志*, e-book.

Table 6: 田金生，"对我县重点人口管理工作的调查与思考," *北京人民警察学院学报*, no. 59 (1999): 27; *铜陵年鉴 1991*, 75, 165; *南岸区年鉴 1993–1997*, 13, 79; *迁安年鉴 1998–1999*, 20, 88; *新野年鉴 1998*, 171; *中国公安年鉴 2000*, 409; *耒阳年鉴 1993*, 41, 180; *耒阳年鉴 1995*, 53, 120; "湖南新宁司法篇," http://city.sina.com.cn/city/t/2011-08-17/170221463.html; *湖南年鉴 1993*, 543; *湖南年鉴 1997*, 371; *平陆年鉴 1995*, 10, 67; *会同年鉴 1992–1995*, 215; *通城年鉴 1995*, 67, 135; *通城年鉴 1996*, 20, 61; *黄石年鉴 1996*, 90, 176; *枣庄年鉴 1997*, 51, 126; *沁水县志 1986–2003*, 484; *黎城县志 1991–2003*, 470; *安义年鉴 1993–1998*, 138, 304; *宁都年鉴 1991–1994*, 153; *大连市志:公安志*, 254; *大连统计年鉴 2017*, 94; *昌图年鉴 1997*, 24, 143; *富锦市志*, 514, 652; *齐齐哈尔市建华区志 1996–2005*, 473, 627–633; *大庆市萨尔图区志 1986–2005*, 410–412; *柞水年鉴 1998–2002*, 124, 253; *商南县志 1991–2010*, 116, 603; *龙岩市志 1988–2002*, 1215; *永安年鉴 1991–1994*, 47, 221; *王建幸*，"浅论派出所基础工作," *公安理论与实践*, no. 3 (1995): 3; *崇明年鉴 1994*, 290; *崇明年鉴 1996*, 261–262; *崇明年鉴 1998*, 223–224; *崇明年鉴 2000*, 201–202; *玛纳斯县志 1986–2010*, 117, 849; *田东年鉴 1994–1998*, 192–193; *来宾年鉴 1991–2000*, 285; *象州年鉴 1991–1995*, 61, 119; *徐州年鉴 1998*, 51, 118; *宜兴年鉴 1992*, 3, 99; *淮阴年鉴 1997*, 91; *双流年鉴 1994*, 51, 95; *大竹年鉴 1993*, 2, 87; *大竹年鉴 1995*, 3, 78; *成都成华志 1990–2005*, 77–78, 231; *临江年鉴 1994–1995*, 141, 225; *保山地区年鉴 1996*, 190, 252; *杭州市人民公安志*, 212; *慈溪市公安志*, 239; *余姚公安志*, 159, 164; *缙云县公安志*, 170, 178.

Table 7: *苍南年鉴 2002*, 93, 151; *北海年鉴 2001–2002*, 46, 100; *定安年鉴 2009*, 83; *张掖综合年鉴 2004–2005*, 9, 686; *湖南新宁司法篇*, http://city.sina.com.cn/city/t/2011-08-17/170221463.html; *湖南年鉴 2005*, 354; *沁水县志*, 484; *灵宝市志 1988–2000*, 651–652; *龙岩市志 1988–2002*, 1215; *龙岩年鉴 2003*, 21; *德化年鉴 2009*, 12, 103; *涵江年鉴 2010*, 9, 174; *芜湖县志 1990–2003*, 462; *启东年鉴 2001*, 93, 191; *新浦年鉴 2002*, 66, 110; *通州年鉴 2006*, 118; *灌南年鉴 2001*, 134; *淮安年鉴 2001*, 37, 101; *长宁年鉴 2010*, 38, 142; *宜宾年鉴 2007*, 26, 112; *通川年鉴 2008*, 129; *通川年鉴 2006*, 338 (2005 population used); *三台年鉴 2002*, 2, 129; *成都年鉴 2007*, 2, 118; *成都成华志 1990–2005*, 77–78, 231; *成华年鉴 2009*, 86; *锦江年鉴 2008*, 87; *锦江年鉴 2009*, 1, 68; *锦江年鉴 2010*, 1, 60; *即墨年鉴 2010*, 1, 97; *平原县志*, 1986–2008, 87, 491; *文登年鉴* 2007–2009, 284–285; *东营区年鉴* 2006, 142; *泰安年鉴 2002*, 262; *南康年鉴 2004*, 43, 116; *弋阳年鉴 2006–2009*, 44, 182; *新建县志 1985–2002*, 115, 667; *呼兰年鉴 2008–2009*, 109–110, 281, 342; *大兴安岭年鉴 2002*, 20, 114; *伊春市年鉴 2006*, 12, 87; *富锦市志*, 514, 652; *齐齐哈尔市建华区志* 1996–2005, 351; *齐齐哈尔年鉴* 2004, 354; *大庆市萨尔图区志 1986–2005*, 410–412; *兴安盟志*I, 332, 1671; *准格尔年鉴 2006*, 63, 159; *乌海年鉴 2000–2001*, 24, 121; *昌吉年鉴 2010*, 30, 163; *云南年鉴 2009*, 39, 124; *大关年鉴 2005*, 74, 159; *大关年鉴 2006*, 89, 186; *大关年鉴 2009*, 67, 191; *泸西年鉴 2005*, 97, 211; *威信年鉴 2001*, 38, 120; *威信年鉴 2004–2005*, 29, 174; *罗平年鉴 2002*, 156; *思茅年鉴 2003*, 136; *云南年鉴 2003*, 380; *鲁甸年鉴 2009*, 265; *渭南年鉴* 2007, 2, 153; *柞水年鉴 1998–2002*,

1, 124; *西乡年鉴 2001–2003*, 223, 268; *南郑年鉴 2004–2005*, 20, 334; *银川年鉴 2006*, 32, 287; *银川年鉴 2008*, 52, 274; *华龙年鉴 2010*, 109, 125; *恩平年鉴 2004–2006*, 131, 231; *本溪年鉴 2002*, 73, 150; *本溪年鉴 2007*, 61, 166; *朝阳年鉴 2004*, 1, 130; *平泉年鉴 2006*, 144, 329; *柳河年鉴 2010*, 136.

Table 8: *普安年鉴 2017*, 226, 297; *普安年鉴 2018*, 134, 429; *普安年鉴 2020*, 147, 464; *瓮安年鉴 2013*, 33, 123; *瓮安年鉴 2015*, 176–177; *黔东南年鉴 2014*, 42, 172; *黔东南年鉴 2018*, 212; *松桃年鉴 2013*, 7, 470; *贵阳年鉴 2018*, 191; *南明年鉴 2019*, 42, 122; *晴隆年鉴 2017*, 99, 101; *通城年鉴 2015*, 14, 70; *黑河年鉴 2014*, 43, 144; *黑河年鉴 2018* 31, 119; *榆中年鉴 2012–2014*, 115, 143; *金昌年鉴 2011*, 301, 316; *金昌年鉴 2012*, 323, 343; *正宁年鉴 2013*, 41, 189; *乌什年鉴 2011*, 112, *乌什年鉴 2012*, 111; *乌什年鉴 2015*, 44, 122; *伊吾年鉴 2011*, 15; *哈密年鉴 2014*, 3, 113; *塔城年鉴 2015*, 97; *长宁年鉴 2012*, 4, 91; *德化年鉴 2011*, 11, 99; *康定年鉴 2019*, 52, 136; *锦江年鉴 2012*, 2, 125; *锦江年鉴 2019*, 9, 250; *游仙年鉴 2016*, 28, 92; *绵竹年鉴 2015*, 39, 97; *灞桥年鉴 2013*, 57, 210; *乌拉特后旗年鉴 2016*, 226, 418; *和平区年鉴 2011*, 3, 283; *和平区年鉴 2012*, 181, 351; *垦利年鉴 2015*, 152; *无棣年鉴 2012–2014*, 45, 176; *巨野年鉴 2018*, 23, 114; *宝丰年鉴 2011*, 102, 243; *虞城年鉴 2013*, 16,124; *芜湖年鉴 2012*, 93; *芜湖年鉴 2013*, 65; *芜湖年鉴 2014*, 72; *芜湖年鉴 2015*, 68; *芜湖年鉴 2016*, 76; *芜湖年鉴 2017*, 18, 82; *普洱年鉴 2015*, 131, 394; *苍南年鉴 2013*, 8, 127.

Table 9: *安龙年鉴 2017*, 207; *普洱年鉴 2015*, 131; *德化年鉴 2011*, 99; *芜湖年鉴 2012*, 93; *芜湖年鉴 2013*, 65; *芜湖年鉴 2014*, 72; *芜湖年鉴 2015*, 68; *芜湖年鉴 2016*, 76; *芜湖年鉴 2017*, 83; *乌特拉后年鉴 2016*, 236; *黔东南年鉴 2014*, 172; *黔东南年鉴 2018*, 212; *黑河年鉴 2014*, 144; *巴彦年鉴 2011*, 169; *瓮安年鉴 2013*, 123; *瓮安年鉴 2015*, 177; *晴隆年鉴 2017*, 101; *涟水年鉴 2003*, 117; *成都市成华区志 1991–2001*, 235; *和平区年鉴 2011*, 183; *内江市市中区年鉴 2016*, 180.

Notes

Introduction

1 Jon Russell, "China's CCTV Surveillance Network Took Just Seven Minutes to Capture BBC Reporter," TechCrunch, December 14, 2017, https://techcrunch.com/2017/12/13/china-cctv-bbc-reporter.

2 Paul Mozur and Allan Krolik, "A Surveillance Net Blankets China's Cities," *New York Times,* December 17, 2019.

3 Josh Chin and Clément Bürge, "Twelve Days in Xinjiang: How China's Surveillance State Overwhelms Daily Life," *Wall Street Journal,* December 17, 2017.

4 Paul Mozur, Claire Fu, and Amy Chang Chien, "How China's Police Used Phones and Faces to Track Protesters," *New York Times,* December 2, 2022.

5 George Soros, "Remarks Delivered at the World Economic Forum," Davos, Switzerland, January 24, 2019, https://www.georgesoros.com/2019/01/24/remarks-delivered-at-the-world-economic-forum-2/.

6 Suzanne Scoggins, *Policing China* (Ithaca, NY: Cornell University Press, 2021); Susan Trevaskes, *Policing Serious Crime in China: From 'Strike Hard' to 'Kill Fewer'* (London: Routledge, 2012); Michael Dutton, *Policing and Punishment in China: From Patriarchy to 'the People'* (Cambridge: Cambridge University Press, 1992); Michael Dutton, *Policing Chinese Politics* (Durham, NC: Duke University Press, 2005); Murray Scot Tanner and Eric Green, "Principals and Secret Agents: Central versus Local Control over Policing and Obstacles to 'Rule of Law' in China," *China Quarterly* 191 (2007): 644–670; Ivan Sun and Yuning Wu, "Chinese Policing in a Time of Transition, 1978–2008," *Journal of Contemporary Criminal Justice* 26, no. 1 (2010): 20–35; Edward Schwarck, "Intelligence and Informatization: The Rise of the Ministry of Public Security in Intelligence Work in China," *China Journal* 80, no. 1 (2018): 1–23.

7 Yuhua Wang and Carl Minzner, "The Rise of the Chinese Security State," *China Quarterly* 222 (2015): 339–359; Yuhua Wang, "Empowering the Police: How the Chinese Communist Party Manages Its Coercive Leaders,"

China Quarterly 219 (2014): 625–648; Xie Yue, "Rising Central Spending on Public Security and the Dilemma Facing Grassroots Officials in China," *Journal of Current Chinese Affairs* 42, no. 2 (2013): 79–109; Sheena Greitens, "Rethinking China's Coercive Capacity: An Examination of PRC Domestic Security Spending, 1992–2012," *China Quarterly* 232 (2017): 1002–1025; Xuezhi Guo, *China's Security State: Philosophy, Evolution, and Politics* (New York: Cambridge University Press, 2012).

8 Sheena Greitens, Myunghee Lee, and Emir Yazici, "Counterterrorism and Preventive Repression: China's Changing Strategy in Xinjiang," *International Security* 44, no. 3 (2020): 9–47; James Tong, *Revenge of the Forbidden City: The Suppression of the Falungong in China, 1999–2005* (New York: Oxford University Press, 2009); Sarah Biddulph, *Legal Reform and Administrative Detention Powers in China* (New York: Cambridge University Press, 2007); Youyu Xu and Hua Ze, eds., *In the Shadow of the Rising Dragon: Stories of Repression in the New China* (New York: Macmillan, 2013); Christian Göbel, "The Political Logic of Protest Repression in China," *Journal of Contemporary China* 30, no. 128 (2021): 169–185; Xiaojun Yan, "Engineering Stability: Authoritarian Political Control over University Students in Post-Deng China," *China Quarterly* 218 (2014): 493–513; Zhou Kai and Xiaojun Yan, "The Quest for Stability: Policing Popular Protest in the People's Republic of China," *Problems of Post-Communism* 60, no. 3 (2014): 3–17.

9 Margaret Roberts, *Censored* (Princeton: Princeton University Press, 2018); Gary King, Jennifer Pan, and Margaret E. Roberts, "How Censorship in China Allows Government Criticism but Silences Collective Expression," *American Political Science Review* 107, no. 2 (2013): 326–343; Rogier Creemers, "Cyber China: Upgrading Propaganda, Public Opinion Work and Social Management for the Twenty-First Century," *Journal of Contemporary China* 26, no. 103 (2017): 85–100; Wen-Hsuan Tsai, "How 'Networked Authoritarianism' Was Operationalized in China: Methods and Procedures of Public Opinion Control," *Journal of Contemporary China* 25, no. 101 (2016): 731–744; H. Christoph Steinhardt, "Defending Stability under Threat: Sensitive Periods and the Repression of Protest in Urban China," *Journal of Contemporary China* 30, no. 130 (2021): 526–549; Rory Truex, "Focal Points, Dissident Calendars, and Preemptive Repression," *Journal of Conflict Resolution* 63, no. 4 (2019): 1032–1052; Xiaojun Yan, "Patrolling Harmony: Pre-emptive Authoritarianism and the Preservation of Stability in W County," *Journal of Contemporary China* 25, no. 99 (2016): 406–421; Yanhua Deng and Kevin O'Brien, "Relational Repression in China: Using Social Ties to Demobilize Protesters," *China Quarterly* 215 (2013): 533–552; Kevin O'Brien and Yanhua

Deng, "Preventing Protest One Person at a Time: Psychological Coercion and Relational Repression in China," *China Review* 17, no. 2 (2017): 179–201; Xi Chen, "Origins of Informal Coercion in China," *Politics & Society* 45, no. 1 (2017): 67–89; Lynette Ong, *Outsourcing Repression: Everyday State Power in Contemporary China* (New York: Oxford University Press, 2022); Martin Dimitrov, *Dictatorship and Information: Authoritarian Regime Resilience in Communist Europe and China* (New York: Oxford University Press, 2022); Jennifer Pan, *Welfare for Autocrats: How Social Assistance in China Cares for Its Rulers* (New York: Oxford University Press, 2020); Daniel Mattingly, *The Art of Political Control in China* (New York: Cambridge University Press, 2020), 154–180.

10 Journalistic accounts include Kai Strittmatter, *We Have Been Harmonized: Life in China's Surveillance State* (New York: Custom House, 2021); and the more substantive and informative work Josh Chin and Liza Lin, *Surveillance State: Inside China's Quest to Launch a New Era of Social Control* (New York: St. Martin's Press, 2022). Valuable scholarly accounts include James Leibold, "Surveillance in China's Xinjiang Region: Ethnic Sorting, Coercion, and Inducement," *Journal of Contemporary China* 29, no. 121 (2020): 46–60; Jessica Batke and Mareike Ohlberg, "State of Surveillance: Government Documents Reveal New Evidence on China's Efforts to Monitor Its People," Asia Society, *ChinaFile,* October 30, 2020, https://www.chinafile.com/state-surveillance-china. Works on the social credit system include Fan Liang, Vishnupriya Das, Nadiya Kostyuk, and Muzammil M. Hussain, "Constructing a Data-Driven Society: China's Social Credit System as a State Surveillance Infrastructure," *Policy & Internet* 10, no. 4 (2018): 415–453; Xu Xu, "To Repress or to Co-opt? Authoritarian Control in the Age of Digital Surveillance," *American Journal of Political Science* 65, no. 2 (2021): 309–325.

11 One work that does answer broader questions is Michael Schoenhals, *Spying for the People: Mao's Secret Agents, 1949–1967* (New York: Cambridge University Press, 2013), but it does not cover the post-Mao period.

12 Christian Davenport, "State Repression and Political Order," *Annual Review of Political Science* 10, no. 1 (2007): 1–23; Mai Hassan, Daniel Mattingly, and Elizabeth R. Nugent, "Political Control," *Annual Review of Political Science* 25, no. 1 (2022): 155–174; Alexander Dallin and George Breslauer, *Political Terror in Communist Systems* (Stanford, CA: Stanford University Press, 1970).

13 Christian Davenport, Hank Johnston, and Carol Mueller, eds., *Repression and Mobilization* (Minneapolis: University of Minnesota Press, 2005); Dursun Peksen and A. Cooper Drury, "Economic Sanctions and Political

Repression: Assessing the Impact of Coercive Diplomacy on Political Freedoms," *Human Rights Review* 10, no. 3 (2009): 393–411.

14 Ronald Wintrobe, *The Political Economy of Dictatorship* (New York: Cambridge University Press, 2000), 342.

15 Robert Barro, "Democracy and Growth," *Journal of Economic Growth* 26, no. 3 (1996): 1–27; Baizhu Chen and Yi Feng, "Some Political Determinants of Economic Growth: Theory and Empirical Implications," *European Journal of Political Economy* 12, no. 4 (1996): 609–627.

16 Alberto Alesina, Sule Özler, Nouriel Roubini, and Phillip Swagel, "Political Instability and Economic Growth," *Journal of Economic Growth* 1, no. 2 (1996): 189–211; Ari Aisen and Francisco José Veiga, "How Does Political Instability Affect Economic Growth?" *European Journal of Political Economy* 29 (2013): 151–167.

17 Sheena Greitens, *Dictators and Their Secret Police: Coercive Institutions and State Violence* (New York: Cambridge University Press, 2016), 12; Jack Paine, "Reframing the Guardianship Dilemma: How the Military's Dual Disloyalty Options Imperil Dictators," *American Political Science Review* 116, no. 4 (2022): 1–18.

18 Andrea Kendall-Taylor and Erica Frantz, "How Autocracies Fall," *Washington Quarterly* 37, no. 1 (2014): 35–47, 37.

19 Michael Makara, "Coup-proofing, Military Defection, and the Arab Spring," *Democracy and Security* 9, no. 4 (2013): 334–359; Ulrich Pilster and Tobias Böhmelt, "Coup-proofing and Military Effectiveness in Interstate Wars, 1967–99," *Conflict Management and Peace Science* 28, no. 4 (2011): 331–350; Greitens, *Dictators and Their Secret Police,* 30–61.

20 Bruce Bueno de Mesquita and Alastair Smith, *The Dictator's Handbook: Why Bad Behavior Is Almost Always Good Politics* (New York: Public Affairs, 2011); Erica Frantz and Andrea Kendall-Taylor, "A Dictator's Toolkit: Understanding How Co-optation Affects Repression in Autocracies," *Journal of Peace Research* 51, no. 3 (2014): 332–346; Johannes Gerschewski, "The Three Pillars of Stability: Legitimation, Repression, and Co-optation in Autocratic Regimes," *Democratization* 20, no. 1 (2013): 13–38.

21 Bruce Gilley, *The Right to Rule: How States Win and Lose Legitimacy* (New York: Columbia University Press, 2009).

22 Zheng Wang, *Never Forget National Humiliation: Historical Memory in Chinese Politics and Foreign Relations* (New York: Columbia University Press, 2014); Suisheng Zhao, *A Nation-State by Construction: Dynamics of Modern Chinese Nationalism* (Stanford, CA: Stanford University Press, 2004); Sergei Guriev and Daniel Treisman, *Spin Dictators: The Changing Face of Tyranny in the 21st Century* (Princeton, NJ: Princeton University Press, 2022).

23 Chappell Lawson, "Mexico's Unfinished Transition: Democratization and Authoritarian Enclaves in Mexico," *Mexican Studies* 16, no. 2 (2000): 267–287; Bruce Dickson, "Integrating Wealth and Power in China: The Communist Party's Embrace of the Private Sector," *China Quarterly* 192 (2007): 827–854.

24 Steven Levitsky and Lucan Way, *Competitive Authoritarianism: Hybrid Regimes after the Cold War* (New York: Cambridge University Press, 2010); Jennifer Gandhi and Adam Przeworski, "Authoritarian Institutions and the Survival of Autocrats," *Comparative Political Studies* 40, no. 11 (2007): 1279–1301.

25 Tiberiu Dragu and Adam Przeworski, "Preventive Repression: Two Types of Moral Hazard," *American Political Science Review* 113, no. 1 (2019): 77–87; Nathan Danneman and Emily Ritter, "Contagious Rebellion and Preemptive Repression," *Journal of Conflict Resolution* 58, no. 2 (2014): 254–279.

26 Kris De Jaegher and Britta Hoyer, "Preemptive Repression: Deterrence, Backfiring, Iron Fists, and Velvet Gloves," *Journal of Conflict Resolution* 63, no. 2 (2019): 502–527.

27 E. K. Bramstedt, *Dictatorship and Political Police* (London: Kegan Paul, 1945); Molly Pucci, *The Secret Police in Communist Eastern Europe* (New Haven, CT: Yale University Press, 2020); Jonathan Adelman, ed., *Terror and Communist Politics: The Role of the Secret Police in Communist States* (Boulder, CO: Westview Press, 1984); Pablo Policzer, *The Rise and Fall of Repression in Chile* (South Bend, IN: University of Notre Dame Press, 2009); Carl Wege, "Iranian Intelligence Organizations," *International Journal of Intelligence and Counter Intelligence* 10, no. 3 (1997): 287–298.

28 Dragu and Przeworski, "Preventive Repression"; Yevgenia Albats, *The State within a State: The KGB and Its Hold on Russia* (New York: Farrar, Straus, Giroux, 1994); Greitens, *Dictators and Their Secret Police.* Examples of powerful spy chiefs include Lavrentiy Beria in the former USSR, Heinrich Himmler in Nazi Germany, and Erich Mielke, the long-serving head of the Stasi in East Germany.

29 Wege, "Iranian Intelligence Organizations," 288; Chile: National Intelligence Directorate (DINA) and National Information Center (CNI), Federation of American Scientists, Intelligence Resource Program, September 11, 1998, https://irp.fas.org/world/chile/dina.htm.

30 Mike Dennis, *The Stasi: Myth and Reality* (London: Longman, 2003), 78–79; Gary Bruce, *The Firm: The Inside Story of the Stasi* (Oxford: Oxford University Press, 2010), 11, 13. The East German state had a population of 16.4 million in 1989.

31 "Citizen Spies," *Deutsche Welle (DW)* Global Media Forum, March 11, 2008, https://www.dw.com/en/east-german-stasi-had-189000-informers-study-says/a-3184486-1.

32 沈晓洪 等, "基层公安机关警力配置现状与思考," *Journal of Jiangxi Public Security College,* no. 141 (July 2010): 107.

33 Joseph Sassoon, *Saddam Hussein's Ba'th Party: Inside an Authoritarian Regime* (New York: Cambridge University Press, 2012), 125.

34 Dennis, *The Stasi*, 99.

35 Bruce, *The Firm*, 80–105.

36 Roderic Camp, *Politics in Mexico: The Decline of Authoritarianism* (Oxford: Oxford University Press, 1999); Harold Crouch, *Government and Society in Malaysia* (Ithaca, NY: Cornell University Press, 1996); Diane Mauzy and Robert Stephen Milne, *Singapore Politics under the People's Action Party* (London: Routledge, 2002).

37 Houchang Chehabi and Juan Linz, eds., *Sultanistic Regimes* (Baltimore, MD: Johns Hopkins University Press, 1998); Gholam Reza Afkhami, *The Life and Times of the Shah* (Berkeley: University of California Press, 2009); David Nicholls, "Haiti: The Rise and Fall of Duvalierism," *Third World Quarterly* 8, no. 4 (1986): 1239–1252; Gary Hawes, *The Philippine State and the Marcos Regime* (Ithaca, NY: Cornell University Press, 1987); Harold Crouch, "Patrimonialism and Military Rule in Indonesia," *World Politics* 31, no. 4 (1979): 571–587.

38 Owen Sirrs, *The Egyptian Intelligence Service: A History of the Mukhabarat, 1910–2009* (London: Routledge, 2010).

39 Pucci, *The Secret Police*; Amy Knight, *The KGB: Police and Politics in the Soviet Union* (Boston, MA: Unwin Hyman, 1988); Schoenhals, *Spying for the People*.

40 Barbara Geddes, Erica Frantz, and Joseph G. Wright, "Military Rule," *Annual Review of Political Science* 17, no. 1 (2014): 147–162.

41 Saddam Hussein's Ba'th Party is generally considered a one-party regime that was better organized than other dictatorships. But the Ba'th Party had a shallow footprint in society and a looser organizational structure than a typical Leninist party. Sassoon, *Saddam Hussein's Ba'th Party*; Thomas Rigby, *Communist Party Membership in the USSR* (Princeton, NJ: Princeton University Press, 2019).

42 Carl Friedrich and Zbigniew Brzezinski, "The General Characteristics of Totalitarian Dictatorship," in *Comparative Government,* ed. Jean Blondel, 3–22 (London: Palgrave, 1969); Juan Linz, *Totalitarian and Authoritarian Regimes* (Boulder, CO: Lynne Rienner, 2000).

43 Kenneth Lieberthal, *Governing China: From Revolution through Reform* (New York: Norton, 2003); Richard Burks, *Dynamics of Communism in Eastern Europe* (Princeton, NJ: Princeton University Press, 2015).

44 "Citizen Spies," *Deutsche Welle (DW);* 吉林省志: 司法公安志, ch. 7 and ch. 9, e-book; 浙江通志: 公安志, 292; 浙江省人口志, Table 3-4-1, e-book; 福建公安志, http://data.fjdsfzw.org.cn/2016-09-21/content_295.html; 湖北公安志, 62, 194; 江西公安志, 60, 80; "化工本14党支部开展社区治安巡逻活动," https://www.cup.edu.cn/chem/dqgz/zthd/151707.htm; "组织在职党员、在职团员、中学生团员社区安全巡逻活动的通知," https://www.bit.edu.cn/tzgg17/qttz/a176500.htm; "西峰区各乡镇召开全面加强社会治安管理工作动员会," http://www.qyswzfw.gov.cn/Show/326262.

45 F. W. Mott, *Imperial China 900–1800* (Cambridge, MA: Harvard University Press, 1999), 140, 753.

46 Mott, *Imperial China 900–1800,* 138–144; Yuchen Song, "Rethinking on Wang Anshi's Reform from Economics Perspective," in *Proceedings of the 2022 3rd International Conference on Language, Art and Cultural Exchange (ICLACE 2022),* 336–341 (Atlantis Press, 2002).

47 Timothy Brook, *The Chinese State in Ming Society* (London: Routledge, 2005), 36.

48 Mott, *Imperial China 900–1800,* 918–919.

49 Mo Tian, "The *Baojia* System as Institutional Control in Manchukuo under Japanese Rule (1932–45)," *Journal of the Economic and Social History of the Orient* 59, no. 4 (2016): 531–554; Ching-Chih Chen, "The Japanese Adaptation of the pao-chia System in Taiwan, 1895–1945," *Journal of Asian Studies* 34, no. 2 (1975): 391–416; Lane Harris, "From Democracy to Bureaucracy: The Baojia in Nationalist Thought and Practice, 1927–1949," *Frontiers of History in China* 8, no. 4 (2013): 517–557.

50 "最新统计数据显示: 党员 9671.2万名 基层党组织 493.6万个," http://www.gov.cn/xinwen/2022-06/29/content_5698405.htm.

51 Seymour Martin Lipset, "Some Social Requisites of Democracy: Economic Development and Political Legitimacy," *American Political Science Review* 53, no. 1 (1959): 69–105; Samuel Huntington, *The Third Wave: Democratization in the Late Twentieth Century* (Norman: University of Oklahoma Press, 1993); Robert Dahl, *Polyarchy: Participation and Opposition* (New Haven: Yale University Press, 1971).

52 Jie Chen and Bruce Dickson, *Allies of the State: China's Private Entrepreneurs and Democratic Change* (Cambridge, MA: Harvard University Press, 2010); Sebastian Heilmann and Elizabeth Perry, eds., *Mao's Invisible Hand: The*

Political Foundations of Adaptive Governance in China (Cambridge, MA: Harvard University Asia Center, 2011); David Shambaugh, *China's Communist Party: Atrophy and Adaptation* (Berkeley: University of California Press, 2008); Andrew Nathan, "Authoritarian Resilience," *Journal of Democracy* 14, no. 1 (2003): 6–17; Teresa Wright, *Accepting Authoritarianism: State-Society Relations in China's Reform Era* (Stanford, CA: Stanford University Press, 2010).

53 See also Yuhua Wang, "Coercive Capacity and the Durability of the Chinese Communist State," *Communist and Post-Communist Studies* 47, no. 1 (2014): 13–25; Yan, "Patrolling Harmony."

54 Joshua Rosenzweig, "Political Prisoners in China: Trends and Implications for US Policy," Testimony to the Congressional-Executive Committee on China, August 3, 2010; Ware Fong, "Depoliticization, Politicization, and Criminalization: How China Has Been Handling Political Prisoners since 1980s," *Journal of Chinese Political Science* 24, no. 2 (2019): 315–339.

55 Wealthy autocracies have stronger repressive capacity. Vincenzo Bove, Jean-Philippe Platteau, and Petros G. Sekeris, "Political Repression in Autocratic Regimes," *Journal of Comparative Economics* 45, no. 2 (2017): 410–428; Michael Ross, "The Political Economy of the Resource Curse," *World Politics* 51, no. 2 (1999): 297–322.

56 Guillermo O'Donnell, Philippe C. Schmitter, and Laurence Whitehead, eds., *Transitions from Authoritarian Rule,* vol. 4, *Tentative Conclusions about Uncertain Democracies* (Baltimore, MD: Johns Hopkins University Press, 1986).

57 Minxin Pei, *China's Trapped Transition: The Limits of Developmental Autocracy* (Cambridge, MA: Harvard University Press, 2006).

58 Adam Przeworski and Fernando Limongi, "Modernization: Theories and Facts," *World Politics* 49, no. 2 (1997): 155–183.

1. The Evolution of the Chinese Surveillance State

1 "管制反革命分子暂行办法," http://www.ce.cn/xwzx/gnsz/szyw/200705/29/t20070529_11526416.shtml; *天津通志：公安志*, 560; *浙江人民公安志*, 254–257.

2 For discussion of mass terror in revolutionary regimes generally, see Steven Levitsky and Lucan Way, *Revolution and Dictatorship: The Violent Origins of Durable Authoritarianism* (Princeton, NJ: Princeton University Press, 2022); on China's terror campaigns specifically, see Frank Dikötter, *Tragedy of Liberation* (New York: Bloomsbury Press, 2013); Yang Kuisong, "Reconsidering the Campaign to Suppress Counterrevolutionaries," *China Quarterly* 193 (March 2008): 102–112.

3 *Statistical Yearbook of China 1990* (Beijing: China Statistical Publishing Co., 1991), 89.

4 *建国以来公安工作大事要览*（北京：群众出版社,2003),21,24, 25, and 36 (hereafter, *公安工作大事要览*).

5 *浙江人民公安志*, 83.

6 *江西公安志*, 186.

7 *上海公安志*, Part 2, Ch. 3, e-book.

8 *江西公安志*: 186; *上海公安志*, Part 2, Ch. 2, e-book; *浙江人民公安志*, 107–108.

9 *公安工作大事要览*, 87–89.

10 *公安工作大事要览*, 137.

11 *公安工作大事要览*, 164–166.

12 "罗瑞卿同志在全国公安厅局长座谈会上的总结发言," in *公安会议文件汇编* (1949.10–1957.9), 529–535. Jiangxi police established files on the "counterrevolutionary social base." *江西公安志*, 72.

13 *公安工作大事要览*.

14 This number includes 171,000 policemen, or *minjing,* who did not have any official rank. In addition, there were 125,000 officers and soldiers in the People's Armed Police. *公安工作大事要览*, 157.

15 *公安工作大事要览*, 344, 564.

16 *Statistical Yearbook of China 1999,* Table 4-1; and *Statistical Yearbook of China 2011,* Table 3-1, e-book.

17 *浙江通志：公安志*, 63.

18 *甘肃省公安志*, 84.

19 Calculations based on *贵州省志：公安志*, 127–129.

20 *象山县公安志*, 290; *鄞县公安志*, 125–128.

21 Michael Schoenhals, *Spying for the People: Mao's Secret Agents, 1949–1967* (New York: Cambridge University Press, 2013), 206, 228.

22 "罗瑞卿在第六次全国公安会议上的报告," May 17, 1957, 173.

23 陈一新, "1955 年公安工作的基本总结和1956年的工作任务（草稿)," *第11次（湖北）省公安会议文件*, 325.

24 "罗瑞卿在第六次全国公安会议上的报告," 185–186.

25 Schoenhals, *Spying for the People,* 1; *公安工作大事要览*, 319.

26 *公安工作大事要览*, 346, 349.

27 *陕西省志:公安志*, 393; *景德镇市志第一卷公安志*, 399–400; "治安保卫委员会暂行组织条例," http://www.gov.cn/zhengce/2020-12/25/content_5573171.htm.

28 The share of CCP and Youth League members was 54 percent in Nanjing in 1963 and 46 percent in Jiaxing in 1960. *南京公安志*, 159; *嘉兴人民公安志*, 297.

29 *吉林省志：司法公安志*, Ch. 7 and Ch. 9, e-book; *浙江通志：公安志*, 292; 浙江省人口志, Table 3-4-1, e-book; *福建公安志*, http://data.fjdsfzw.org.cn/2016-09-21/content_295.html; *湖北公安志*, 62, 194; *江西公安志*, 60, 80.

30 *中国法律年鉴1987–1997（珍藏版）*（北京:中国法律出版社,1998), 739, 742, 754–755, 765, 767, 778, 780, 795, 797, 811, 827, and 829.

31 *公安工作大事要览*, 152.

32 *公安工作大事要览*, 165.

33 *公安工作大事要览*, 507.

34 *来宾政法志*, 41–43, 56.

35 Michael Dutton, "Policing the Chinese Household: A Comparison of Modern and Ancient Forms," *Economy and Society* 17, no. 2 (1988): 195–224; Wai-po Huen, "Household Registration System in the Qing Dynasty: Precursor to the PRC's Hukou System," *China Report* 32, no. 4 (1996): 395–418.

36 Zhang Qingwu, "Basic Facts on the Household Registration System," ed. and trans. Michael Dutton, *Chinese Economic Studies* 22 (1988): 3–106; Michael Dutton, *Policing and Punishment in China: From Patriarchy to the People* (New York: Cambridge University Press, 1992); Frederic Wakeman Jr., *Policing Shanghai, 1927–1937* (Berkeley: University of California Press, 1995).

37 Fei-Ling Wang, *China's Hukou System: Organizing through Division and Exclusion* (Stanford, CA: Stanford University Press, 2005); Kam Wing Chan and Li Zhang, "The Hukou System and Rural-Urban Migration in China: Processes and Changes," *China Quarterly* 160 (1999): 818–855; Tiejun Cheng and Mark Selden, "The Origins and Social Consequences of China's Hukou System," *China Quarterly* 139 (1994): 644–668.

38 *Statistical Yearbook of China 1999,* Table 4-1, e-book.

39 The MPS issued its "Interim Rules of Urban Hukou Management" ("城市户口管理暂行条例") in July 1951. http://www.hljcourt.gov.cn/lawdb/show.php?fid=48&key=%D0%D0%D5%FE;*公安工作大事要览*, 63.

40 *江西公安志*, 72.

41 *湖南省志：政法志*, Ch. 2, e-book.

42 "中华人民共和国户口登记条例," https://www.waizi.org.cn/doc/120275.html.

43 *公安工作大事要览*, 143–144.

44 *浙江人民公安志*, 290–294.

45 *浙江人民公安志*, 290–294.

46 *广东公安志*, 209–211; *湖南省志: ：政法志*, Ch. 2, e-book.

47 *江西公安志*, 73; *湖南省志：政法志*, Ch. 2, e-book; *浙江人民公安志*, 290–294; *广东公安志*, 209–211.

48 The Chinese approach to surveillance also seems to imitate the Stalinist practice of cataloging the population. See David Shearer, *Policing Stalin's Socialism* (New Haven: Yale University Press, 2009), 158–180.

49 *陕西省志：公安志*, 723; *公安工作大事要览*, 164–166.

50 *江西公安志*, 49; *天津通志：公安志*, 550, 560; *甘肃省志：公安志*, 466–467.

51 *公安工作大事要览*, 419, 573.

52 *中国法律年鉴 1987–1997（珍藏版）*, 161. The 20 million-plus figure is also mentioned in a news dispatch cited in the MPS's *Major Events in Public Security Work 1949–2000*. *公安工作大事要览*, 573.

53 *上海公安志*, Ch. 3, e-book; *浙江人民公安志*, 257–258; *贵州省志：公安志*, 581.

54 *浙江人民公安志*, 257; *甘肃省公安志*, 464.

55 *浙江人民公安志*, 257–258; *湖南省志：政法志*, e-book; *江西公安志*, 49; *福建公安志*, http://data.fjdsfzw.org.cn/2016-09-21/content_295.html.

56 *广西公安志*，http://lib.gxdfz.org.cn/view-a63-220.html;*上海公安志*, Ch. 3, e-book; *贵州省志：公安志*, 580.

57 *绍兴市志*, http://www.sx.gov.cn/col/col1462780/index.html.

58 *长兴公安志*, 387; *岱山县公安志*, 158; *舟山市公安志*, 210.

59 *杭州市人民公安志*, 210; *绍兴县公安志*, 142.

60 "关于重点人口管理工作的暂行规定." Reference to this regulation is contained in *公安工作大事要览*, 100–101.

61 "城市治安管理工作细则," *山东省国情网*, "重点人口管理."

62 *黑龙江省志-公安志*, 373; population for 1959 obtained from https://m.gotohui.com/pdata-0/1959.

63 *黑龙江省志-公安志*, 373; https://m.gotohui.com/pdata-0/1959; *重庆市志：公安志*, Ch. 7, e-book; *杭州市人民公安志*, 210; *建德市公安志*, 191, 260; *余杭公安志*, 79, 343; *岱山县公安志*, 158, 165; *天台县公安志*, 98, 100; *宁海县公安志*, 337; *慈溪市公安志*, 237, 251.

64 "关于加强政法工作的指示," http://cpc.people.com.cn/GB/4519165.html#.

65 *公安工作大事要览*, 344, 564.

66 *中国法律年鉴1987–1997（珍藏版）*, 739, 778.

67 湖北公安志, 259.

68 *公安工作大事要览*, 448.

69 Local political-legal committees had very little staff. In some counties, they had only two or three people, including the secretary. In other

counties, the committees had no full-time staff or secretary. *祥云县志 1978–2005,* 741; *元谋县志 1978–2005,* 207; *河源市源城区志 1988–2003,* 684.

70 *公安工作大事要览,* 507–508, 528, 537.

71 "关于加强国家安全部和公安部合作的意见," *公安工作大事要览,* 596.

72 *公安工作大事要览,* 576–577.

73 Ruan Chongwu, a liberal, was minister of the MPS between August 1985 and March 1987.

74 *公安工作大事要览,* 557, 646, 699, 731.

75 Revenue data from *Statistical Yearbook of China 2021,* Table 7-1, e-book.

76 Murray Scot Tanner, "State Coercion and the Balance of Awe: The 1983–1986 'Stern Blows' Anti-crime Campaign," *China Journal* 44 (2000): 93–125.

77 *公安工作大事要览,* 653.

78 Susan Trevaskes, "Severe and Swift Justice in China," *British Journal of Criminology* 47, no. 1 (2007): 23–41.

79 Merle Goldman, *Sowing the Seeds of Democracy in China: Political Reform in the Deng Xiaoping Era* (Cambridge, MA: Harvard University Press, 1994).

80 Net urban population growth in the 1990s was 157 million (including births). *Statistical Yearbook of China 2001,* Table 4.1, e-book; Xiaobo Lu and Elizabeth Perry, eds., *Danwei: The Changing Chinese Workplace in Historical and Comparative Perspective* (Armonk, NY: M. E. Sharpe, 1997).

81 Timothy Brook and B. Michael Frolic, *Civil Society in China* (London: Routledge, 2015).

82 Murray Scot Tanner, "China Rethinks Unrest," *Washington Quarterly* 27, no. 3 (2004): 137–156.

83 "中共中央关于维护社会稳定加强政法工作的通知," http://www.reformdata.org/1990/0402/4106.shtml.

84 "关于加强公安工作的决定," *公安工作大事要览,* 872, 875.

85 Yuhua Wang and Carl Minzner, "The Rise of the Chinese Security State," *China Quarterly* 222 (2015): 339–359.

86 公安工作大事要览, 820, 917, 932, 1084–1085, 1132, 1141.

87 "中共中央、国务院关于加强社会治安综合治理的决定," http://www.reformdata.org/1991/0219/4159.shtml; "中共中央国务院关于进一步加强社会治安综合治理的意见," http://www.gov.cn/gongbao/content/2001/content_61190.htm.

88 Hualing Fu, "Zhou Yongkang and the Recent Police Reform in China," *Australian & New Zealand Journal of Criminology* 38, no. 2 (2005): 241–253.

89 "中共中央关于进一步加强和改进公安工作的决定，" http://www.reformdata.org/2003/1118/4921.shtml.

90 "中共中央关于进一步加强和改进公安工作的决定"; Yuhua Wang, "Empowering the Police: How the Chinese Communist Party Manages Its Coercive Leaders," *China Quarterly* 219 (2014): 625–648.

91 Sheena Greitens, "Rethinking China's Coercive Capacity: An Examination of PRC Domestic-Security Spending, 1992–2012," *China Quarterly* 232 (2017): 1002–1025.

92 *Statistical Yearbook of China 2020,* Table 7-1, e-book.

93 Since 2012, the government has released only the total amount of domestic security spending, which includes spending on the People's Armed Police. Because the share of spending on the PAP averaged about 16 percent between 2002 and 2011, it is reasonable to assume that about 84 percent of total domestic security expenditures between 2012 and 2020 was devoted to public security, the procuratorate (that is, public prosecutors), and the courts.

94 The consumer price index rose 472 percent between 1991 and 2020. *Statistical Yearbook of China 1999,* Table 9-2; *Statistical Yearbook of China 2021,* Table 5-2, e-book.

95 *地方财政统计资料 1995*（北京：新华出版社, 1996), 314, 316; *地方财政统计资料 1996*（北京：中国财政经济出版社, 1998), 317, 319.

96 *中国法律年鉴1987–1997（珍藏版）*, 778; 沈晓洪 等, "基层公安机关警力配置现状与思考," *Journal of Jiangxi Public Security College,* no. 141 (July 2010): 107.

97 *湖北公安志*, 259.

98 *浙江通志：公安志*, 63.

99 *贵州省志：公安志*, 129.

100 *湖北公安志*, 259.

101 *浙江通志：公安志*, 63. The estimate of 10,000 additional police nationwide is a conservative one. China has thirty-one provincial jurisdictions. If each gained an average of 350 police officers—fewer than the number in Zhejiang and Hubei—the net increase would exceed 10,000.

102 *公安工作大事要览*, 880.

103 *公安工作大事要览*, 988, 1073–1074, 1227.

104 "中共中央关于维护社会稳定加强政法工作的通知."

105 "维稳办走上前台," *双周刊*, no. 8 (2009): 44–46.

106 Weishan Miao and Wei Lei, "Policy Review: The Cyberspace Administration of China," *Global Media and Communication* 12, no. 3 (2016): 337–340.

107 Beibei Tang, "Grid Governance in China's Urban Middle-class Neighbourhoods," *China Quarterly* 241 (2020): 43–61; Fan Liang, Vishnupriya Das, Nadiya Kostyuk, and Muzammil M. Hussain, "Constructing a Data-driven Society: China's Social Credit System as a State Surveillance Infrastructure," *Policy & Internet* 10, no. 4 (2018): 415–453.

108 *Statistical Yearbook of China 2021,* Table 7-1, e-book.

109 *Statistical Yearbook of China,* various years.

110 Little is known about this new commission. See Joel Wuthnow, "China's New 'Black Box': Problems and Prospects for the Central National Security Commission," *China Quarterly* 232 (2017): 886–903.

111 "十八大、十九大后落马省部级及以上高官名单," http://district.ce.cn/newarea/sddy/201410/03/t20141003_3638299.shtml.

112 Sheena Greitens, "Surveillance, Security, and Liberal Democracy in the Post-COVID World," *International Organization* 74 (S1) (2020): E169–E190.

113 Yu Sun and Wilfred Wang, "Governing with Health Code: Standardising China's Data Network Systems during Covid-19," *Policy & Internet* 14 (2022): 673–687; Fan Liang, "COVID-19 and Health Code: How Digital Platforms Tackle the Pandemic in China," *Social Media + Society* 6, no. 3 (2020): 1–4.

114 "河南健康码变色之警示," *China Economic Weekly,* June 30, 2022, 52–55; Chris Buckley, Vivian Wang, and Keith Bradsher, "Living by the Code," *New York Times,* January 30, 2022.

115 National Health Commission, "关于印发"十四五"全民健康信息化规划的通知," http://www.nhc.gov.cn/cms-search/xxgk/getManuscriptXxgk.htm?id=49eb570ca79a42f688f9efac42e3c0f1.

116 Minxin Pei, "Grid Management: China's Latest Institutional Tool of Social Control," *China Leadership Monitor* 67 (Spring 2021), https://www.prcleader.org/pei-grid-management.

117 习近平, "在中央政治局常委会会议研究应对新型冠状病毒肺炎疫情工作时的讲话," https://www.12371.cn/2022/09/03/ARTI1662190489127492.shtml.

118 "先锋街道网格化推进老年人疫苗接种," http://www.tongzhou.gov.cn/tzzt/fkxd/content/192aff3a-0796-4609-b3da-46e2edd0d509.html; "网格员深入社区轮岗值守," http://www.szlhq.gov.cn/jdbxxgkml/mzjdb/dtxx_124615/gzdt_124616/content/post_10108548.html; "广东省全面推行

"网格化"疫情防控," http://www.gov.cn/xinwen/2020-02/11/content_5477195.htm; "广灵县网格化管理筑牢疫情防控网," http://www.dt.gov.cn/dtzww/xqxx1/202205/1b8681d9cd2641588a78339b9b9416d1.shtml.

119 "习近平指示强调：把'枫桥经验'坚持好、发展好," http://www.gov.cn/ldhd/2013-10/11/content_2504878.htm.

Table 1.1 data sources: *广东公安志*, 180; *新中国五十五年统计资料汇编*（北京：中国统计出版社，2005), 708; *陕西省志：公安志*, 725; *跨世纪的中国人口 陕西卷* (北京：中国统计出版社, 1991): 20–22; *天津通志：公安志*, 550, 562; *江西公安志*, 49–50, 80–81; *湖南省志：政法志*, Ch. 3, e-book; *湖南省志：人口志*, e-book; *上海公安志* , Ch. 8, e-book; *上海通志，第三卷*, e-book; *福建公安志*, http://data.fjdsfzw.org.cn/2016-09-21/content_295.html; *甘肃省公安志*, 400, 465; *吉林省志：司法公安志*, Ch. 10, e-book; *浙江人民公安志*, 256–260; *广西公安志*, http://lib.gxdfz.org.cn/view-a63-220.html; *广西通志 (1979–2005)*, http://www.gxdfz.org.cn/flbg/szgx/201710/t20171013_47365.html; *贵州省志：公安志*, 580–581; *贵州省志：人口和计划生育*, 3.

Table 1.2 data sources: *公检法支出财务统计资料，1991–1995* (南京：江苏科技出版社，1991), 166, 169; *中国统计年鉴*, various years (北京：中国统计年鉴出版社).

2. Command, Control, and Coordination

1 Sheena Greitens, *Dictators and Their Secret Police: Coercive Institutions and State Violence* (New York: Cambridge University Press, 2016), 12; Blake McMahon and Branislav Slantchev, "The Guardianship Dilemma: Regime Security through and from the Armed Forces," *American Political Science Review* 109, no. 2 (2015): 297–313.

2 Ephraim Kahana and Muhammad Suwaed, *Historical Dictionary of Middle Eastern Intelligence* (Lanham, MD: Scarecrow Press, 2009), 66, 210–211.

3 "Manuel Contreras, Head of Chile's Spy Agency under Pinochet, Dies Aged 86," *Guardian*, August 8, 2015.

4 Mike Dennis, *The Stasi: Myth and Reality* (Harlow: Pearson/Longman, 2003), 40–46.

5 Murray Scot Tanner and Eric Green, "Principals and Secret Agents: Central versus Local Control over Policing and Obstacles to 'Rule of Law' in China," *China Quarterly* 191 (2007): 644–670.

6 This friction between local and central authorities is a symptom of "fragmented authoritarianism." See Kenneth Lieberthal and David Lampton,

eds., *Bureaucracy, Politics, and Decision Making in Post-Mao China* (Berkeley: University of California Press, 2018).

7 The Chinese names of these commissions are: 中央国家安全委员会, 中央全面深化改革委员会, 中央网络安全和信息化委员会, 中央外事工作委员会, 中央军民融合发展委员会.

8 "全国社会治安综合治理工作大事记," 法制日报, September 25, 2001, http://news.sina.com.cn/c/2001-09-25/365090.html. In 2011, the party changed the name of this commission to "中央社会管理综合治理委员会," but in 2014, the party changed the name back to its original. "中央综治委恢复"治安"原名," *澎湃新闻*, October 10, 2014, http://m.thepaper.cn/kuaibao_detail.jsp?contid=1270571&from=kuaibao.

9 Wen-Hsuan Tsai and Wang Zhou, "Integrated Fragmentation and the Role of Leading Small Groups in Chinese Politics," *China Journal,* no. 82 (July 2019): 1–22; Alice Miller, "More Already on the Central Committee's Leading Small Groups," *China Leadership Monitor,* no. 44, Hoover Institution, July 28, 2014, https://www.hoover.org/research/more-already-central-committees-leading-small-groups.

10 Their Chinese names are 中央防范和处理邪教问题领导小组 and 中央维护稳定领导小组.

11 Sarah Cook and Leeshai Lemish, "The 610 Office: Policing the Chinese Spirit," *China Brief* 11, no. 17, Jamestown Foundation, September 16, 2011, https://jamestown.org/program/the-610-office-policing-the-chinese-spirit; "刘金国不再担任610办公室主任," *中国经济网*, http://www.xinhuanet.com/politics/2015-05/26/c_127843253.htm.

12 James Tong, *Revenge of the Forbidden City: The Suppression of the Falungong in China, 1999–2005* (New York: Oxford University Press, 2009).

13 "维稳办走上前台," *双周 刊*, no. 8 (2009): 44.

14 "中央维稳办调研组来我市调研," *莱芜日报*, July 16, 2014, 2; "中央维稳办调研组来我市," *济宁日报*, August 30, 2009, 1; "中央维稳办调研组来抚调研," *抚州日报*, May 14, 2015, 1.

15 "告诉你一个完整的中央新疆工作协调小组," *澎湃新闻*, September 10, 2014, https://www.thepaper.cn/newsDetail_forward_1267650; "汪洋出席第七次全国对口支援新疆工作会议," 北京青年报, July 16, 2019, https://news.sina.com.cn/c/2019-07-16/doc-ihytcerm4115379.shtml.

16 "西藏工作协调小组工作范围扩至4省藏区," 南方都市报, August 19, 2010, http://news.sina.com.cn/c/2010-08-19/055420928594.shtml.

17 "平安中国建设协调小组成立," *澎湃新闻*, April 22, 2020, https://www.thepaper.cn/newsDetail_forward_7083492.

18 “平安中国建设协调小组四个专项组亮相,” *新京报*, July 28, 2020, https://www.sohu.com/a/410223007_114988.

19 The Chinese name of the document is “中共中央进一步加强和改进公安工作的决定.”

20 While Hu Jintao did not attend these conferences, he did meet with a select group of delegates in 2007. “总书记出席政法工作会议,” http://news.takungpao.com/mainland/focus/2014-01/2164561.html.

21 For example, the PLCs in Xiangyun County and Yuanmo County, Yunnan, had only skeletal staffs of two to three people. *祥云县志* 1978–2005, 741; *元谋县志* 1978–2005, 207.

22 The CCP used the same model when creating the China Cyber Administration and its local outfits. But there is a crucial difference between the local offices of the cyber administration and the local PLCs: the cyber administration offices are part of the local CCP propaganda department, while the PLCs are stand-alone party bureaucracies that enjoy a higher political status.

23 *公安工作大事要览*, 964.

24 In the Chinese system, municipalities—a status held by about 300 large cities—and prefectures occupy the same administrative rank. All, including so-called autonomous prefectures, are at the second rank of the administrative hierarchy, below provinces, of which they are subdivisions.

25 周永坤, “论党委政法委员会之改革,” *法学*, no. 5 (2012): 3–13; 刘忠, “政法委的构成与运作,” *环球法律评论*, no. 3 (2017): 16–38.

26 “中共中央关于加强政法工作的指示,” http://www.71.cn/2011/0930/632692.shtml.

27 刘忠, “政法委的构成与运作,” 34.

28 “中共中央关于维护社会稳定加强政法工作的通知,” http://www.reformdata.org/1990/0402/4106.shtml.

29 In his speech to delegates at the December 1995 political-legal conference, Jiang Zemin declared that the size and funding of the local PLCs would be increased. *公安工作大事要览*, 1047–1048; 李永浩等, “全面推进依法治国背景下中央政法委改革探讨,” *淮阴师范学院学报*, no. 3 (2016): 301–308.

30 殷家国, “基层政法委在履行职责中存在的主要问题及对策,” *贵州省政法管理干部学院学报*, no. 1 (1996): 37–38.

31 “中共中央关于进一步加强和改进公安工作的决定,” http://www.reformdata.org/2003/1118/4921.shtml; “中共中央关于进一步加强和改进公安工作的决定”; Yuhua Wang and Carl Minzner, “The Rise of the Chinese Security State,” *China Quarterly* 222 (2015): 339–359; Yuhua

Wang, "Empowering the Police: How the Chinese Communist Party Manages Its Coercive Leaders," *China Quarterly* 219 (2014): 625–648.

32 钟金燕, "中共政法委制度的历史考察," *中共党史研究*, no. 4 (2014): 116–124; 曾林妙, 陈科霖, "中国国家治理中的政法委制度," *国家治理评论* (April 2020): 5–15.

33 David Lampton, "Xi Jinping and the National Security Commission: Policy Coordination and Political Power," *Journal of Contemporary China* 24, no. 95 (2015): 759–777.

34 For analysis of the purge, see Guoguang Wu, "Continuous Purges: Xi's Control of the Public Security Apparatus and the Changing Dynamics of CCP Elite Politics," *China Leadership Monitor,* no. 66 (Winter 2020), https://www.prcleader.org/wu; Sheena Greitens, "The *Saohei* Campaign, Protection Umbrellas, and China's Changing Political-Legal Apparatus," *China Leadership Monitor,* no. 65 (Fall 2020), https://www.prcleader.org/greitens-1.

35 Chen Yixin claimed that the political-legal sector counted 2.7 million total personnel. But China has about 2 million uniformed police alone, suggesting that this number likely does not include the People's Armed Police—a force estimated at 1.5 million strong. 中央政法委, "政法队伍整顿教育，全国12576名干警投案," https://www.infzm.com/contents/207864; "全国政法队伍教育整顿领导小组：16个中央督导组近期到位," *今日观察新闻社*, March 25, 2021, http://newsaum.com/fztd/1096.html.

36 "中共中央印发中国共产党政法工作条例," http://www.gov.cn/zhengce/2019-01/18/content_5359135.htm.

37 钟金燕，"中共政法委制度的历史考察," 121–123.

38 *元谋年鉴* 1996, 109; *元谋年鉴*2011, 189.

39 "中共武汉市委政法委员会概况," http://www.wuhan.gov.cn/ztzl/yjs/2020_52695/drz/zgwhswzfwyh_36901/sndjs_36903/202110/t20211015_1795862.shtml.

40 The PLC of Chenghua District, Chengdu, had a staff of twenty-two in 2021. "2022 年中共成都市成华区委政法委 部门预算," http://www.chenghua.gov.cn/chqrmzfw/c144318cgt/2022-01/20/37f3d52743a449b7a27c18982b211743/files/c358be787d1f4d74ba596c2be9c96d41.pdf. The PLC of Neihuang County, Henan Province, with a population of nearly 800,000 in 2018, had a staff of twenty-seven in the same year. *内黄年鉴* 2019, 37, 190.

41 *开封市禹王台区年鉴* 2019, 25, 66; *天长年鉴* 2020, 62.

42 *北京密云年鉴*2019, 1, 112.

43 The Chinese Wikipedia page on the CPLC lists eleven departments bearing titles that sound potentially accurate. However, no official sources confirm the existence of these departments. https://zh.wikipedia.org/wiki/中共中央政法委员会.

44 贵州政法委, http://www.guizhou.gov.cn/ztzl/gzsczzjxxgkzl_1794/sygbmhdwczyjsjsgjf/zggzswzfwyh/201609/t20160905_436885.html.

45 Hangzhou's PLC provides a good example. 杭州市政法委, https://www.hangzhou.gov.cn/col/col809713/index.html.

46 For illustrative descriptions of the organization of various local PLCs, see *北京密云年鉴* 2019, 112; *内黄年鉴* 2019, 190; *维西傈僳族自治县年鉴* 2019, 148; *黄龙年鉴* 2015, 6.

47 *深圳政法年鉴* 2000.

48 Many local yearbooks contain brief summaries of the PLCs' accomplishments. More detailed summaries can be found in the section on PLCs in *中共邯郸年鉴* (北京：中共党史出版社) published between 2002 and 2019.

49 For more on strict security measures applied surrounding sensitive dates, see Rory Truex, "Focal Points, Dissident Calendars, and Preemptive Repression," *Journal of Conflict Resolution* 63, no. 4 (2019): 1032–1052.

50 *中共天津工作* 2017 (天津：天津人民出版社, 2018).

51 *崇明年鉴* 2013, 60.

52 *深圳政法年鉴* 1997, 27; *深圳政法年鉴* 1998, 2; *芜湖年鉴* 2010, 234.

53 *深圳政法年鉴* 2000 (深圳：海天出版社, 2001), 161; *崇明年鉴* 2012, 62; *崇明年鉴* 2013, 58–59; *芜湖年鉴* 2011, 228.

54 *齐齐哈尔年鉴* 2017, 21; *米东年鉴* *2006*, 202; *芜湖年鉴* 2011, 228–229.

55 福田区政法委, "政法委第4 周工作汇总."

3. Organizing Surveillance

1 沈晓洪 等，"基层公安机关警力配置现状与思考," *Journal of Jiangxi Public Security College,* no. 141 (July 2010): 107.

2 Sheena Greitens, "Rethinking China's Coercive Capacity: An Examination of PRC Domestic Security Spending, 1992–2012," *China Quarterly* 232 (2017): 1002–1025.

3 Nearly all dictatorships, including the communist regimes of the former Eastern Bloc, have an interior ministry in charge of regular policing and a secret police agency responsible for spying and surveillance of dissidents.

4 The minister of public security has the rank of a vice premier and often serves on the Secretariat of the Central Committee, whereas the minister of state security is merely a member of the Central Committee. One MSS minister was promoted to the Politburo—Chen Wenqing, in 2022—but he immediately relinquished his MSS position.

5 The Central Investigation Department was established as the result of Zhou's proposal to reorganize the party's intelligence apparatus in 1955. 宋月红, "鲜为人知的 '中央调查部,'" https://history.ifeng.com/c/81u9hOywRFt.

6 毛泽东，"镇压反革命必须实行党的群众路线," *毛泽东文集 第6卷* (北京：人民出版社出版, 1999), 161–162.

7 *建国以来公安大事要览*, 5; 沈志华，*苏联专家在中国, 1948–1960*（北京：社会科学文献出版社, 2015).

8 *新疆通志*, vol. 20, 194. Post-2019 name changes are reflected in some local yearbooks; for example, *盱眙年鉴 2020*, 37; *淮安年鉴 2020*, 287.

9 Gao Guangjun, interview with author via Zoom, February 12, 2022.

10 By comparison, there were about 110,000 police officers in the "economic" sector. "罗瑞卿在第六次全国公安会议上的总结," June 17, 1957, in *公安会议文件汇编* (1949.10–1957.9), 208.

11 *新疆通志*, vol. 20, 194; *新疆年鉴 1986*, 30.

12 刘新民，"当前制约政保工作诸因素简析," *河南公安学刊*, no. 21 (1995): 44.

13 *株洲公安志*, 74; *大庆市萨尔图区志*, 392; *枣阳年鉴 2012–2013*, 315; *水磨沟年鉴 2013*, 180; *武汉公安年鉴 2015*, 27, 41.

14 Gary Bruce, *The Firm: The Inside Story of the Stasi* (Oxford: Oxford University Press, 2010), 11, 13.

15 许晓明，刘英武，"国内安全保卫工作改革初探," *公安研究*, no. 107 (2003): 75.

16 苏全霖，刘黎明，"论国内安全保卫工作法治化," *Journal of Shanxi Police Academy* 24, no. 1 (2016): 44.

17 刘新民，"当前制约政保工作诸因素简析" 4: 42.

18 "公安部一局大学处张伟处长来校调研," http://gac.snnu.edu.cn/info/1022/1177.htm; "省市公安部门来我校调研维族学生管理工作," https://www.hznu.edu.cn/c/2011-09-09/327492.shtml; "北京市公安局文保支队来校检查会商校园安稳工作," https://news.bistu.edu.cn/zhxw/202009/t20200925_222599.html.

19 "滨州市公安局国保支队来学院调研指导工作," http://www.sdbky.cn/info/1003/1213.htm; "郑州市高校维稳安保工作考核组来我院检查指导工作," http://www.hagmc.edu.cn/info/1100/2482.htm.

20 "郑州市高校维稳安保工作考核组来我院检查指导工作;" "公安部一局大学处张伟处长来校调研;" "省市公安部门来我校调研维族学生管理工作."

21 *西安年鉴 2010*, 164; *西安年鉴 2012*, 159–160; *西安年鉴 2013*, 143; *西安年鉴 2014*, 94.

22 *武汉公安年鉴 2010*, 152–154; *武汉公安年鉴 2012*, 181–184.

23 Anonymous Shanghai academic in exile, interview with author, May 26, 2019, Princeton, NJ; Xu Youyu, interview with author, May 25, 2019, Flushing, NY; Teng Biao, interview with author, March 4, 2020, Claremont, CA.

24 谷福生等编, *新时期公安派出所工作全书* (北京：中国人民公安大学出版社, 2005), 542–543.

25 The Chinese term *shehuihua* refers to the recruitment of spies from all social strata. "石城县公安局主要职责（内设机构职能)," http://www.shicheng.gov.cn/xxgk/xxgkml/glgk/jgzn/28/t28_1124823.html.

26 谷福生等编, *新时期公安派出所工作全书*, 547–548.

27 *株洲公安志*, 74–78.

28 *大庆市萨尔图区 1986–2005*, 392–393.

29 *舒兰市志 1986–2002*, 305–306.

30 *汉源年鉴 2017*, 144; *武汉公安年鉴 2010*, 94, *武汉公安年鉴 2014*, 67; *大庆市萨尔图区志 1986–2005*, 392–393; *米易年鉴 2011*, 87; *磐石市志 1991–2003*, 351–352.

31 *汉源年鉴 2017*, 144; *武汉公安年鉴 2010*, 94; *武汉公安年鉴 2014*, 67.

32 *西双版纳年鉴 2011*, 243; *米易年鉴 2009*, 72.

33 The document is titled "关于开展国内安全保卫工作对象基础调查的意见." *北京公安年鉴 2003*, 121.

34 *株洲公安志*, 74–78.

35 *武汉公安年鉴 2006*, 76.

36 *北京公安年鉴 2001*, 122–123.

37 *北京公安年鉴 2006*, 76–77.

38 谷福生等编, *新时期公安派出所工作全书*, 543; 张先福 等编, *公安派出所窗口服务与执法执勤工作规范* (北京：群众出版社, 2006), 93–95.

39 *大庆市萨尔图区志 1986–2005*, 392.

40 *磐石市志 1991–2003*, 351–352; *西双版纳年鉴 2011*, 243; *米易年鉴 2009*, 83.

41 *株洲公安志*, 74–78.

42 *舒兰市志1986–2002*, 305–306; *米易年鉴 2009*, 83.

43 *磐石市志 1991–2003*, 351–352.

44 The Chinese title is "派出所国内安全保卫工作规范."

45 *北京公安年鉴 2001*, 122–123; *岳阳楼区年鉴 2017*, 175; *四川乐山市公安局文件汇编* 3, 13.

46 *北京公安年鉴 2006*, 77; *内黄年鉴 2018*, 205.

47 *瓮安年鉴 2013*, 123.

48 *磐石市志 1991–2003*, 351–352; *福州市台江区志 1991–2005*, 432–433; *磐石市志 1991–2003*, 351–352.

49 *武汉公安年鉴 2014*, 67.

50 *磐石市志 1991–2003*, 351–352; *岳阳楼区年鉴 2017*, 175.

51 "公安派出所组织条例," *中国人权年鉴*（北京：当代世界出版社, 2000), 325.

52 魏琳，"新时期公安基层派出所职能定位探讨," *Journal of Sichuan Police College* 29, no. 3 (2017): 99–100; 鲍遂献, "杭州会议与公安派出所改革," *Journal of Jiangxi Public Security College* (November 2002): 8.

53 "公安部为派出所招兵买马 充实七万警力到基层," *瞭望东方周刊,* http://news.sohu.com/20050622/n226046035.shtml.

54 Xinhua, "公安部：2013 年社会治安平稳," http://politics.people.com.cn/n/2014/0103/c70731-24019741.html.

55 "广东省公安厅关于加强新时代公安派出所建设工作的意见," https://www.fnhg.net/xuesheng/fyebcql1am.html.

56 河北日报，"省公安厅：年底前派出所警力不低于总警力40%," cpc.people.com.cn/n/2013/0921/c87228-22980265.html.

57 中华人民共和国公安部, "公安派出所执法执勤工作规范," https://new.qq.com/omn/20211201/20211201A0B4YY00.html.

58 杨玉章 ed., *金水公安改革之路*（北京：中国人民公安大学出版社, 2003), 354.

59 *北京公安年鉴 2002*, 121; *四川乐山市公安局文件汇编* 3, 30.

60 *开封禹王台区年鉴 2019*, 70; *开封禹王台区年鉴 2018*, 103–104.

61 *卫东区年鉴 2019*, 152–153; *卫东区年鉴 2020*, 170–174; *峨山年鉴 2018*, 56–57.

62 *和龙年鉴 2018*, 128; *鹰潭年鉴 2018*, 189; *文峰区年鉴 2019*, 166.

63 *开封禹王台区年鉴 2017*, 83; *开封禹王台区年鉴 2020*, 91–92, 94.

64 *观澜年鉴 2018*, 65.

65 Wan Yanhai, interview with author, May 25, 2019, Flushing, NY; Teng Biao interview, March 4, 2020; Wang Tiancheng, interview with author, May 26, 2019, Princeton, NJ; anonymous Chinese Academy of Social Sciences researcher in exile, interview with author, May 26, 2019, Princeton, NJ.

66 Teng Biao interview, March 4, 2020.

67 Hua Ze, interview with author, May 26, 2019, Princeton, NJ.

68 Wang Tiancheng interview, May 26, 2019; Wan Yanhai interview, May 25, 2019; anonymous Chinese Academy of Social Sciences researcher in exile interview, May 26, 2019.

69 "公安派出所档案管理办法," http://www.elinklaw.com/zsglmobile/lawView.aspx?id=15882.

70 *华龙年鉴 2020*, 176; *隆回年鉴 2019*, 176.

71 *峨山年鉴 2018*, 56–57; *卫东区年鉴 2020*, 170–171; *观澜年鉴 2018*, 65.

72 *北京石景山年鉴 2006*, 201.

73 *信阳年鉴 2000*, 155–156; *陕西年鉴 1994*, 51.

74 The organization chart of the MSS can be found in Xuezhi Guo, *China's Security State: Philosophy, Evolution, and Politics* (New York: Cambridge University Press, 2012), 365.

75 *临江年鉴 1996–1997*, 151.

76 *中共邢台年鉴 2003*, 345–346.

77 *信阳年鉴* 2012, 226; *太仓年鉴* 2000, 113.

78 *武汉年鉴 1998*, 90; *汉江年鉴 2019*, 139.

79 *信阳年鉴 2000*, 155–156; *临江年鉴 1996–97*, 151.

80 *大理年鉴* 1999, 208.

81 *鄂州年鉴 1997*, 141; *成都年鉴 1997*, 88.

82 *衡阳年鉴 2003*, 182.

83 *成都年鉴 1997*, 88; *太仓年鉴 2000*, 113; *常熟年鉴 1991–1995*, 122; *曲阜年鉴 1996–1998*, 150.

84 *成都年鉴 1997*, 88; *长沙年鉴 1997*, 86; 曲阜年鉴 *1996–1998*, 150.

85 *迪庆年鉴 2014*, 221–222; *曲阜年鉴 1996–1998*, 150.

86 *迪庆 年鉴 2014*, 221–222.

87 *凉山年鉴 2000*, 111.

88 *喀什年鉴* 2001, 165.

89 *曲阜年鉴 1996–1998*, 150; *即墨年鉴 1992–1998*, 332.

90 *迪庆年鉴 2014*, 221–222; *景德镇年鉴 2000*, 117.

91 *六盘水年鉴 2000*, 252.

92 *凉山年鉴 2000*, 111.

93 *无锡年鉴 2001*, 99.

94 *宜昌年鉴 1999*, 167–168.

95 *六盘水年鉴 2000*, 25; *衡阳年鉴 2003*, 182; *大理年鉴 1999*, 208; *迪庆年鉴 2015*, 237.

96 Teng Biao interview, March 4, 2020.

97 Hua Ze interview, May 26, 2019.

98 *景德镇年鉴 2003*, 128; *景德镇年鉴 1999*, 111; *芜湖年鉴 2004*, 243.

99 Anonymous Zhenzhou NGO activist in exile, interview with author, May 27, 2019, Flushing, NY.

100 Anonymous Beijing academic in exile, interview with author, May 27, 2019, Flushing, NY.

101 Wan Yanhai interview, May 25, 2019.

102 Gao Guangjun interview, February 12, 2022.

4. Spies and Informants

1 Adrian James, *Understanding Police Intelligence Work* (Bristol: Policy Press, 2016); J. Mitchell Miller, "Becoming an Informant," *Justice Quarterly* 28, no. 2 (2011): 203–220; Colin Atkinson, "Mind the Grass! Exploring the Assessment of Informant Coverage in Policing and Law Enforcement," *Journal of Policing, Intelligence and Counter Terrorism* 14, no. 1 (2019): 1–19; Cyrille Fijnaut and Gary Marx, eds., *Undercover: Police Surveillance in Comparative Perspective* (Boston: Kluwer, 1995).

2 David Garrow, "FBI Political Harassment and FBI Historiography: Analyzing Informants and Measuring the Effects," *Public Historian* 10, no. 4 (1988): 5–18.

3 "Citizen Spies," *Deutsche Welle (DW)* Global Media Forum, March 11, 2008, https://www.dw.com/en/east-german-stasi-had-189000-informers-study-says/a-3184486-1; Martin Dimitrov and Joseph Sassoon, "State Security, Information, and Repression," *Journal of Cold War Studies* 16, no. 2 (2014): 3–31, 15; "Population of Bulgaria," retrieved from FRED, Federal Reserve Bank of St. Louis, https://fred.stlouisfed.org/series/POPTTLBGA173NUPN.

4 Xu Xu and Xin Jin, "The Autocratic Roots of Social Distrust," *Journal of Comparative Economics* 46, no. 1 (2018): 362–380.

5 Fan Yang, Shizong Wang, and Zhihan Zhang, "State-enlisted Voluntarism in China: The Role of Public Security Volunteers in Social Stability Maintenance," *China Quarterly* 249 (2022): 47–67.

6 Michael Schoenhals, *Spying for the People: Mao's Secret Agents, 1949–1967* (New York: Cambridge University Press, 2013).

7 The Chinese title of the document is "刑事特情工作细则;" 邓立军, "中国现代秘密侦查史稽考," *四川警察学院学报*, no. 3 (2014): 1–6.

8 The 2001 version is "刑事特情工作规定." The two other regulations are "缉毒特情工作管理试用办法" and "狱内侦查工作细则;" 韦洁雯, "浅谈我国刑事特情侦查," *知识一力量*, October 2017, http://www.chinaqking.com/yc/2018/1031601.html.

9 The "Detailed Rules" are not publicly available. This passage is cited in 陈肯, "论刑事特情制度的构建," *法制与社会*，6 (2012): 36.

10 程雷，"特情侦查立法问题研究," *Criminal Law Review*, no. 2 (2011): 515–537.

11 高光俊，"如何识别中共特情," https://sites.google.com/site/mybooksindex/17-M-C-H-Commie-1979-2011/gao-guang-jun-cp-agent.

12 "三门县公安局刑警大队岗位职责," http://www.sanmen.gov.cn/art/2010/12/1/art_1229319970_3161927.html.

13 林建设，“推进刑事特情工作'四化'建设,” *Journal of Wuhan Public Security Cadres College* 87 (2009): 66; 陈玉凡，“新形势下的刑事特情工作,” *Journal of Henan Police College,* 29 (1996): 22; *灞桥年鉴 2019*, 235; *齐河年鉴 2017*, 176.

14 *陕西省志:公安志*, 557.

15 *灞桥年鉴 2019*, 235.

16 The process for hiring teqing is laid out in “秘密力量建设方案的规划,” *齐河年鉴2016*, 226.

17 *陕西省志: 公安志*, 563.

18 陈玉凡，“新形势下的刑事特情工作”: 20–21.

19 On spy tenures, see 高光俊, “如何识别中共特情.” For long-term recruiting plans, see 齐河年鉴 *2016*, 226; 齐河年鉴 *2017*, 176.

20 “北京市公安局内设机构和所属机构职责,” http://gaj.beijing.gov.cn/zfxxgk/jgzn/202001/t20200102_1554557.html.

21 Gao Guangjun, interview with author via Zoom, February 12, 2022.

22 Shaanxi had a population of 36.9 million in 2003. *Statistical Yearbook of China 2004,* Table 4-3, e-book.

23 霍启兴, “西宁市刑侦队伍和基础工作现状、问题及对策,” *Journal of Liaoning Police Academy* 4 (2007): 54–56.

24 杨玉章 ed., *金水公安改革之路* （北京：中国人民公安大学出版社, 2003), 319; *稻城年鉴 2003–2008*, 184.

25 “三门县公安局刑警大队岗位职责.”

26 Xinhua, “公安部：2013 年社会治安平稳,” http://politics.people.com.cn/n/2014/0103/c70731-24019741.html.

27 程雷, “特情侦查立法问题研究.”

28 刘硕，王文章，“对深入开展刑事特情工作的思考”: 28.

29 陈肯，“论刑事特情制度的构建”:36–37; 刘硕，王文章,“对深入开展刑事特情工作的思考”: 29; 陈玉凡，“新形势下的刑事特情工作”: 21.

30 Gao Guangjun, interview with author via Zoom, February 12, 2022; 吴婷，”基层公安财务管理问题探讨,” *Foreign Investment in China,* no. 262 (2012): 129.

31 The Chinese title of the Xining spy-funding program is “西宁市举报案件线索奖励办法.” 霍启兴, “西宁市刑侦队伍和基础工作现状、问题及对策”: 55; https://news.sina.com.cn/c/2005-11-23/14447519185s.shtml.

32 高光俊, “如何识别中共特情.”

33 Gao Guangjun, interview with author via Zoom, February 12, 2022.

34 *陕西省志：公安志*, 562; “西安将新增千辆出租车,” https://www.chinanews.com.cn/gn/2011/03-04/2883594.shtml.

35 *陕西省志：公安志*, 558.

36 Zhang Lin, interview with author, May 25, 2019, Flushing, NY.

37 Yang Zili, telephone interview with author, April 24, 2022.

38 *武胜年鉴 2007*, 204; *瓮安年鉴 2013*, 123; *瓮安年鉴 2015*, 130; *乐至年鉴 2013*, 181; *乐至年鉴 2014*, 153.

39 *巴彦年鉴 2006*, 226.

40 *齐河年鉴 2017*, 176; *内蒙古阿荣旗公安局文件汇编* 2, 93; *灞桥年鉴 2019*, 235; *尼勒克年鉴 2010–2011*, 72.

41 *巩留年鉴 2014*, 107; *富县年鉴 2018*, 96.

42 *米东区年鉴 2011*, 179.

43 *洛南年鉴 1999*, 289; *泸西年鉴 2004*, 214; *大竹年鉴 2010*, 132.

44 "公安派出所档案管理办法(试行)," http://www.elinklaw.com/zsgl mobile/lawView.aspx?id=15882.

45 The zhi'an ermu approval form is "治安耳目呈批表."

46 The zhi'an ermu termination form is "撤销治安耳目呈批表." 冯文光，张菠, *社区警务实用教程*, 209–213; 张先福 等，*公安派出所窗口服务与执法执勤工作规范* (北京：群众出版社, 2006), 142–144.

47 张先福 等，*公安派出所窗口服务与执法执勤工作规范*, 144.

48 李忠信，*中国社区警务研究*（北京：群众出版社, 1999), 192.

49 张先福 等，*公安派出所窗口服务与执法执勤工作规范*, 141.

50 杨玉章 ed., *金水公安改革之路*, 354.

51 *建德年鉴 2012*, 289.

52 *普陀年鉴 2020*, 103; *樟树年鉴 2009*, 211.

53 *舒兰市志 1986–2002*, 305–306; *磐石市志 1991–2003*, 351.

54 *大庆市萨尔图区志 1986–2005*, 392; *克拉玛依区年鉴 2014*, 96.

55 *金昌年鉴 2011*, 304.

56 *大庆市萨尔图区志 1986–2005*, 393.

57 *岢岚年鉴 2017*, 165.

58 Xu Youyu, interview with author, May 25, 2019, Flushing, NY; anonymous Chinese Academy of Social Sciences researcher in exile, interview with author, May 26, 2019, Princeton, NJ.

59 谢岳，维稳的政治逻辑 (Hong Kong: Tsinghua Bookstore Co., 2013), 104–108.

60 广东政法网, "浅谈义务治安维稳信息员队伍建设," June 23, 2010, http://www.gdzf.org.cn/ztzl/zwzt/jm/201006/t20100623_99912.htm; *章贡年鉴* 2007, 234; *定海年鉴* 2008, 67, 194.

61 "良庆区综治维稳信息工作奖励办法," May 20, 2012, http://www.liangqing.gov.cn/zl/lqz/tzgg/t2291145.html; 道里区政法网, "阿城区委维稳办建立维稳信息'三网络'卓见成效," December 20, 2013, http://acheng.hrbzfw.gov.cn/index/detail/type/25012.html; 西宁市综治办、维稳办, "关于进一步加强基层综治维稳信息员队伍建设及信息奖励工作的通知," August 9, 2013, http://www.chinapeace.gov.cn/chinapeace/c28644/2013-08

/09/content_12079529.shtml; “中共深圳市龙岗区委办公室深圳市龙岗区人民政府办公室龙岗区加强维稳联络员、信息员和人民调解员队伍建设实施办法,” February 15, 2011, http://apps.lg.gov.cn/lgzx/gb02vgh/201511/013696d59c5a472092b00f801381ea96.shtml; “德兴市民众维稳信息员队伍建设工程实施方案,” http://www.zgdx.gov.cn/ttt.asp?id=83187.

62 *海珠年鉴 2009*, 242.

63 *秦淮年鉴 2016*, 279; *贺兰年鉴 2012*, 216; *大关年鉴 2011*, 172.

64 阿城区政法网, “阿城区委维稳办建立维稳信息,” acheng.hrbzfw.gov.cn/index/detail/type/25012.html.

65 *灵川年鉴 2014*, 128; *秦淮年鉴 2016*, 279.

66 新京报, “国防部称中国民兵数已从3000万减至800万,” December 17, 2011, http://www.chinanews.com/gn/2011/12-17/3539161.shtml.

67 *天长年鉴 2014*, 193.

68 *太原市杏花岭年鉴 2014*, 235; *舟山年鉴 2017*, 243; *徐汇年鉴 2017*, 136.

69 *泰和年鉴 2015*, 88; *北川羌族自治县年鉴 2018*, 97; *新密年鉴 2013*, 101.

70 *云安年鉴 2013*, 184; *高安年鉴2013*, 40.

71 北京青年报, “北京10万信息员收集涉恐涉暴情报,” May 30, 2014, http://cpc.people.com.cn/n/2014/0530/c87228-25087026.html; population data from *北京年鉴 2015*, 1.

72 *北京海淀区年鉴 2015*, 126, 425; *北京西城区年鉴 2015*, 161, 414.

Table 4.1 data source: *陕西省志：公安志*, 554–558.

Table 4.2 data sources: *瓮安年鉴 2013*, 123; *瓮安年鉴 2015*, 130; *汉源年鉴 2017*, 144; *北川羌族自治县年鉴 2017*, 105; *齐河年鉴 2015*, 152; *乐至年鉴2014*, 153; *巴彦年鉴 2015*, 128; *富县年鉴2018*, 96; *庄浪年鉴 2018*, 156; *灞桥年鉴 2019*, 235; *云岩年鉴 2020*, 137.

5. Mass Surveillance Programs

1 David Shearer, *Policing Stalin's Socialism: Repression and Social Order in the Soviet Union, 1924–1953* (New Haven, CT: Yale University Press, 2009), 158–180.

2 “公安部重点人口管理工作规定,” https://zhuanlan.zhihu.com/p/441008901.

3 Previously, ex-convicts within five years of release from reeducation through labor (*laojiao*) were also included under the KP program. The punishment of reeducation through labor was eliminated in 2013.

4 Standard KP forms can be found at http://www.inmis.com/rarfile/Pspms_Help/Node13.html.

5 “公安部重点人口管理规定,” August 26, 2020, http://www.jsfw8.com/fw/202005/701558.html. The document does not state the identity of the issuer; however, the names of the police stations referenced indicate that it was issued by the PSB of Zixin, Hunan.

6 *黑龙江省志:公安志*, 375–377.

7 *长春市志: 公安志*, 504.

8 *余杭公安志*, 343–344; *舟山市公安志*, 211; *岱山县公安志*, 159; *建德市公安志*, 194; *象山县公安志*, 183.

9 *齐齐哈尔建华区志 1996–2005*, 351.

10 *灵宝市志 1988–2000*, 651–652.

11 *芜湖县志 1990–2003*, 462.

12 *长兴公安志*, 392; *永康市公安志*, 8; *温岭市公安志*, 146–147; *金华市公安志*, 195; *武义县公安志*, 209; *新昌县公安志*, 179.

13 Sheena Greitens, “Rethinking China's Coercive Capacity,” *China Quarterly* 232 (2017): 1002–1025.

14 罗勤,“重点人口管理的五大问题,” *青少年犯罪问题,* no. 6 (1997): 21–22.

15 侯建军, “关于重点人口社会化管理的几点思考,” *Journal of Fujian Police Academy,* no. 5 (2009): 31–35; 郭峰翔, “当前重点人口失控的原因及对策,” *公安大学学报*, no. 2 (1991): 55–57.

16 郭奕晶, “关于加强和创新重点人口管理工作的思考,” *Journal of Shandong Police College,* no. 2 (2012): 138–143.

17 林立, “构建和谐社会过程中社区人口重点管理模式的探讨,” *Journal of Shanghai Police College* 16, no. 2 (2006): 53–56; 王明媚 “重点人口管理水平提升,” *贵州警官职业学院学报*，no. 6 (2016): 121; 普艳梅, 李长亮 “当前重点人口管理存在问题原因分析,” *云南警官学院学报*, no. 2 (2010): 75–78; 陈建, 胡长海, “浅析新形势下重点人员动态管控对策,” *河南警察学院学报*22, no. 4 (2013): 57–60.

18 A Chinese tech company in the business of tracking KI targets has disclosed that local political-legal committees are in charge of the big data system used in KI monitoring. 中版北斗, “政法大数据重点人员管控系统,” http://www.zbbds.com/view-1005.html.

19 The Chinese name of the platform is 公安部重点人员管控系统, referenced in *四川公安厅情报中心文件汇编* 14, 3; China Digital Times, “重点人员,” https://chinadigitaltimes.net/space/重点人员.

20 “浙江省公安机关重点人员动态管控工作规范 (试行),” https://chinadigitaltimes.net/chinese/127487.html.

21 Emile Dirks and Sarah Cook, “China's Surveillance State Has Tens of Millions of New Targets,” *Foreign Policy,* October 21, 2019.

22 *四川公安厅情报中心文件汇编* 14, 2.

23 *本溪年鉴 2005*, 135.

24 *乌鲁木齐县年鉴 2003*, 129; *米东年鉴 2007*, 211.

25 The Jiuzhaigou PSB provides no details about these categories and subtypes. *九寨沟年鉴 1999–2005*, 297.

26 "湖南资兴市公安局," August 26, 2020, http://www.jsfw8.com/fw/202005/701558.html.

27 *石嘴山年鉴 2018*, 248; *芜湖年鉴 2012*, 93; *瓮安年鉴 2013*, 123.

28 "公安部重点人口管理工作规定."

29 *岱山县公安志*, 160–162.

30 *椒江公安志*, 237; *余杭公安志*, 344; *建德市公安志*, 192.

31 *椒江公安志*, 238.

32 *长兴公安志*, 390.

33 *椒江公安志*, 238.

34 *巴彦年鉴 2015*, 128; *巴彦年鉴 2017*, 184; *巩留年鉴 2014*, 107; *北川羌族自治县年鉴 2017*, 105. The yearbooks do not disclose criteria for categorizing a target within a particular level.

35 "浙江省公安机关重点人员动态管控工作规范-试行."

36 "临时布控工作规范," *四川公安厅情报中心文件汇编*2, 3–7; "贵州省公安机关重点人口动态管控规定," mentioned in六盘水市公安局, "市局治安支队开展重点人口管理业务培训," September 26, 2019, http://gaj.gzlps.gov.cn/gzdt/bmdt/201709/t20170917_12947960.html.

37 "略阳县兴州街道办事处关于涉诈重点人员管控工作的安排意见," http://www.lueyang.gov.cn/lyxzf/lyzwgk/zfwj/gzbwj/202204/640e1c40a7054fabb32f3cb67ca608c8.shtml.

38 *安远年鉴 2014*, 155; *渝水年鉴 2010*, 127.

39 *威县年鉴 2015*, 132; *齐河年鉴 2014*, 158.

40 *北川羌族自治县年鉴 2017*, 105; *巴彦年鉴 2017*, 184; *巩留年鉴 2014*, 107.

41 洪伟, "重点人口管理中的多点应对法," *湖北警官学院学报*, no. 3 (2010): 121; 郭奕晶, "关于加强和创新重点人口管理工作的思考," *Journal of Shandong Police College*, no. 2 (2012): 138–143. James Tong describes surveillance of Falun Gong leaders carried out by local officials. James W. Tong, *Revenge of the Forbidden City: The Suppression of the Falungong in China, 1999–2005* (New York: Oxford University Press, 2009), 66–67.

42 殷文杰, 孙国良, "网格化管理在社会治安防控中的应用研究," *Journal of Hubei University of Police*, no. 12 (2013): 60–61.

43 *福城年鉴 2018*, 103; *西盟年鉴 2018*, 169; *维西傈僳族自治县年鉴 2020*, 287; *朝阳年鉴 2018*, 137.

44 四合派出所, "强化重点人口管理工作," March 9, 2016, http://sihe.gdxf.gov.cn/content/detail/59b8f6d9798d98456c4722d0.html.

45 云浮市云城区人民政府, "云城街土门村委联合辖区派出所认真做好'两会'期间重点人员稳控工作," May 21, 2020, http://www.yfyunchengqu.gov.cn/ycqrmzf/wzdh/zjdt/content/post_1339512.html.

46 *南岸区年鉴 2017*, 256.

47 *北川羌族自治县年鉴 2017*, 105; *道里年鉴 2019*, 125.

48 *瓮安年鉴 2016*, 156; *德江年鉴 2015*, 213.

49 *西盟年鉴 2014*, 123.

50 道真自治县人民政府, "道真自治县公安局机构设置," August 20, 2020, http://www.gzdaozhen.gov.cn/zfbm/gaj/jgsz_5696741.

51 Yang Zili, telephone interview with author, April 24, 2022.

52 *北京信息化年鉴 2010*, 224.

53 *郑州铁路局年鉴 2016*, 322.

54 *汕尾市城区年鉴 2016*, 134.

55 *中卫年鉴 2020*, 154.

56 海盟高科, "产品中心," http://www.goldweb.cn/zdry.

57 Police use a color-coding system indicating five levels of alert, from most to least serious: red, orange, yellow, blue, and white. In Guiyang in 2018, 87 percent of alerts were coded blue. *贵阳年鉴 2019*, 170.

58 Teng Biao, interview with author, March 4, 2020, Claremont, CA.

59 Xu Youyu, interview with author, May 25, 2019, Flushing, NY.

60 Teng Biao interview, March 4, 2020.

61 Wang Tiancheng interview, May 26, 2019, Princeton, NJ.

62 Wang Tiancheng interview, May 26, 2019; Wan Yanhai interview with author, May 25, 2019, Flushing, NY.

63 Anonymous Chinese Academy of Social Sciences researcher in exile, interview with author, May 26, 2019, Princeton, NJ.

64 Xu Youyu interview, May 25, 2019; Hua Ze, interview with author, May 26, 2019, Princeton, NJ.

65 Xu Youyu interview, May 25, 2019; Zhang Lin, interview with author, May 25, 2019, Flushing, NY.

66 Wang Tiancheng interview, May 26, 2019; Wang Qingying, interview with author, May 25, 2019, Flushing, NY.

6. Controlling "Battlefield Positions"

1 Anthony Braga, Andrew Papachristos, and David Hureau, "The Effects of Hot Spots Policing on Crime: An Updated Systematic Review and Meta-analysis," *Justice Quarterly* 31, no. 4 (2014): 633–663; David Weisburd

and Cody Telep, "Hot Spots Policing: What We Know and What We Need to Know," *Journal of Contemporary Criminal Justice* 30, no. 2 (2014): 200–220.

2 In 2000, CCP General Secretary Jiang Zemin designated the Internet a "new important battlefield position" in the ideological struggle with "domestic and external hostile forces." Jiang Zemin, "在中央思想政治工作会议上的讲话," *Selected Works of Jiang Zemin,* vol. 3 (Beijing: Renmin chubanshe, 2006), 94.

3 Jens Gieske, *The History of the Stasi: East Germany's Secret Police, 1945–1990* (New York: Berghahn Books, 2015), 100.

4 上海市公安局徐汇分局，"浅析如何加强阵地控制工作," *Journal of Shanghai Police College,* no. 1 (2007): 165; *金昌年鉴 2010*, 286.

5 *武汉公安年鉴 2009*, 56; *北京石景山年鉴 1997–2005*, 336.

6 马忠红，"侦查阵地控制的困境与出路," *Journal of Guangzhou Police College,* no. 1 (2009): 12–16; 张建平, "论深化基层公安刑侦工作改革," *Journal of Jiangsu Police Officer College* 18, no. 4 (2003): 70.

7 德清县公安局课题组，"加强治安阵地控制的若干思考," *Journal of Zhejiang Police College,* no. 4 (2007): 88.

8 The Chinese name for the system is 特种行业治安管理信息系统.

9 普康迪, "特种行业治安管理信息系统," http://www.chinapcd.com/pages/solution_1.html; 成都川大科鸿新技术研究所, "公安部-特种行业及公共娱乐场所管理系统," https://www.khnt.com/case/4/124.html.

10 *蚌山年鉴 2013*, 243; *芜湖县年鉴 2020*, 130.

11 *咸阳年鉴 2018*, 325.

12 吴明山，廉旭，"论新时期特种行业管理治安管理工作," *公安研究*, no. 100 (2003): 35–38.

13 *章贡年鉴* 2009, 129.

14 王英豪，马晨，"外卖配送行业的阵地控制分析," *法制与社会*, no. 3 (2020): 168–169.

15 郑本人, "关于调整拓宽阵地控制的思考," *Journal of Liaoning Police Academy,* no. 4 (2004): 31–36.

16 *崇明县志 1985–2004*, 481; *攀枝花年鉴 1996*, 151.

17 童永正, "派出所基础建设必须以小基础为突破口," *Journal of Shanghai Police College* 16, no. 4 (2006): 15–20.

18 Enze Han and Christopher Paik, "Dynamics of Political Resistance in Tibet: Religious Repression and Controversies of Demographic Change," *China Quarterly* 217 (2014): 69–98.

19 西藏自治区民宗委，"西藏：建立寺庙管理长效机制," *中国宗教*, no. 7 (2013): 68.

20 甘孜州人大民宗委，"甘孜州藏传佛教事务管理逐步驶入法制化轨道," December 31, 2021, http://www.scspc.gov.cn/mzzjwyh/jyjl_653/201412/t20141230_24679.html; 甘孜藏族自治州人民政府，"甘孜藏族自治

州藏传佛教寺院民主管理委员会班子管理办法（修订),” October 13, 2015, http://www.gzz.gov.cn/gzzrmzf/c100234/201510/a9a737ef305b4a6e958345642dcee18e.shtml.

21 李万虎 “西藏寺庙管理体制改革研究,” *西藏发展论坛*, no. 6 (2015): 62.

22 *昌都年鉴 2015*, 135.

23 “甘孜藏族自治州藏传佛教寺院民主管理委员会班子管理办法（修订).”

24 “甘孜藏族自治州藏传佛教寺院民主管理委员会班子管理办法（修订).”

25 *四川年鉴 2016*, 28; *昌都年鉴 2015*, 181–182.

26 甘孜藏族自治州人民政府, “甘孜藏族自治州宗教活动管理办法,” September 15, 2017, http://www.gzz.gov.cn/gzzrmzf/c100234/201709/3c0d07223a8e4fcaa1d236d8dd8d3aba.shtml.

27 *康定年鉴 2018*, 88.

28 *昌都年鉴 2015*, 181–182.

29 *稻城年鉴 2019*, 90; *稻城年鉴 2016*, 112.

30 *西藏年鉴 2013*, 261; *康定年鉴 2015*, 26–27; *石渠年鉴 2019*, 72.

31 *雅江年鉴 2014*, 63; *稻城年鉴 2016*, 112.

32 *马尔康市年鉴 2019*, 190–191; *康定年鉴2018*, 86.

33 *昌都年鉴 2015*, 181–182; *康定年鉴 2018*, 87; *康定年鉴 2015*, 26–27.

34 *马尔康市年鉴 2019*, 190–191; *西藏年鉴 2011*, 261.

35 郑洲, 马杰华, “加强和创新藏区社会管理研究：以拉萨寺庙管理为例,” *Journal of Ethnology*, no. 17 (2013): 71–74.

36 *康定年鉴 2018*, 88.

37 *康定年鉴 2018*, 88; *稻城年鉴 2019*, 89–90.

38 Liza Lin, Eva Xiao, and Jonathan Cheng, “China Targets Another Region in Ethnic Assimilation,” *Wall Street Journal*, July 16, 2021.

39 *西藏年鉴 2011*, 261.

40 Sergio Rodríguez Tejada, “Surveillance and Student Dissent: The Case of the Franco Dictatorship,” *Surveillance & Society* 12, no. 4 (2014): 528–546; Ricardo Medeiros Pimenta and Lucas Melgaço, “Brazilian Universities under Surveillance: Information Control during the Military Dictatorship, 1964 to 1985,” in *Histories of State Surveillance in Europe and Beyond*, ed. Kees Boersma, Rosamunde van Brakel, Chiara Fonio, and Pieter Wagenaar, 118–131 (London: Routledge, 2014).

41 Xiaojun Yan, “Engineering Stability: Authoritarian Political Control over University Students in Post-Deng China,” *China Quarterly* 128 (2014): 493–513; Stanley Rosen, “The Effect of Post-4 June Re-Education Campaigns on Chinese Students,” *China Quarterly* 134 (1993): 310–334.

42 河南师范大学, "国家教育委员会、公安部 关于进一步加强高等学校内部保卫工作的通知," December 7, 2019, https://www.htu.edu.cn/bwc/2019/0712/c11989a148545/pagem.htm.

43 华东师范大学, "高等学校内部保卫工作规定（试行）," May 7, 2018, http://bwc.ecnu.edu.cn/68/33/c13378a157747/page.htm.

44 The Chinese title of the document is "关于做好抵御境外利用宗教对高校进行渗透和防范校园传教工作的意见."

45 呼和浩特民族学院, "关于抵御境外利用宗教对校园进行渗透和防范校园传教工作的实施意见," https://www.imnc.edu.cn/zzb/info/1024/1283.htm; Changsha Medical University, "长沙医学院防渗透工作实施方案," September 21, 2019, http://part.csmu.edu.cn:82/bwc/index.php?_m=mod_article&_a=article_content&article_id=424. For an example of a university-issued political-security provision, see Huanan Agricultural University's regulation: 华南农业大学, "政治安全责任制度实施办法," October 24, 2018, https://xngk.scau.edu.cn/2019/0311/c2681a162911/page.htm.

46 山东大学, "山大概况," https://www.sdu.edu.cn/sdgk/sdjj.htm; *山东大学年鉴* 2004, 284.

47 *南开大学年鉴 2013*, 186.

48 *合肥工业大学年鉴 2001*, 234; *兰州大学年鉴 2015*, 210.

49 *中国政法大学年鉴 2017*, 202.

50 *江南大学年鉴 2020*, 69; *南开大学年鉴 2013*, 186; *贵州大学年鉴 2011*, 295; *贵州大学年鉴 2017*, 381; *南开大学年鉴 2015*, 190.

51 *浙江教育年鉴 2015*, 110; *江南大学年鉴 2019*, 124; *兰州大学年鉴 2014*, 236; *兰州大学年鉴 2015*, 210.

52 *湖北教育年鉴 2012*, 122.

53 *贵州大学年鉴 2014*, 370; *华侨大学年鉴 2014*, 420.

54 *南开大学年鉴 2015*, 190.

55 *合肥工业大学年鉴 2011*, 334; *合肥工业大学年鉴 2012*, 344.

56 *兰州大学年鉴 2014*, 236; *兰州大学年鉴 2015*, 210.

57 *宁夏大学年鉴 2009*, 109; *宁夏大学年鉴 2011*, 85; *宁夏大学年鉴 2012*, 77; *宁夏大学年鉴 2015*, 64; *大连理工大学年鉴 2016*, 171.

58 *贵州大学年鉴 2016*, 468; *贵州大学年鉴 2018*, 91; *贵州大学年鉴 2019*, 392.

59 *南开大学年鉴 2010*, 171.

60 *中国政法大学年鉴 2012*, 230.

61 *无锡轻工大学年鉴 1993–94*, 189; *无锡轻工大学年鉴 1995*, 185; *无锡轻工大学年鉴 1996*, 153.

62 *合肥工业大学年鉴 2007*, 231; *合肥工业大学年鉴 2008*, 276.

63 *华中科技大学年鉴 2004*, 219; *华中科技大学年鉴 2005*, 225.

64 *江南大学年鉴 2001*, 52; *江南大学年鉴 2003*, 87.

65 *合肥工业大学年鉴 2001*, 235.

66 The title of the Hunan document is "湖南省教育系统维稳综治安全信息报送管理暂行办法"; the title of the Hubei document is "湖北省高校工委、省教育厅关于做好全省教育系统维稳情报信息报送工作的通知." The texts of these two regulations are not available. 湖南工学院, "湖南工学院二级学院维稳综治安全信息报送管理办法," July 1, 2017, http://www.hnit.edu.cn/bwc/info/1013/1243.htm; Wuhan University of Communication, "关于做好我校维稳情报信息报送工作的通知," September 29, 2016, http://www.whmc.edu.cn/dzbgs/dzbgs_sy/show-40005.aspx.

67 Beijing Foreign Studies University, "关于组建北京外国语大学学生舆情信息员队伍的通知," April 9, 2012, https://de.bfsu.edu.cn/info/1041/1594.htm; Central China Normal University Wuhan Communication College, "关于做好我校维稳情报信息报送工作的通知," September 29, 2016, http://www.whmc.edu.cn/dzbgs/dzbgs_sy/show-40005.aspx; Ji'An College, "维稳信息收集上报制度," October 17, 2017, http://www.japt.com.cn/bwc/info/1120/1180.htm; South China Agricultural University, "学生综合信息员工作细则," May 9, 2018, https://life.scau.edu.cn/zsjyw/2018/0509/c10455a106084/page.htm; Changsha Medical University, "维稳信息报送制度暂行规定," http://part.csmu.edu.cn:82/bwc/index.php?_m=mod_article&_a=article_content&article_id=328; Hunan Institute of Technology, "二级学院维稳综治安全信息报送管理办法," July 1, 2019, http://www.hnit.edu.cn/bwc/info/1013/1243.htm.

68 "华南农业大学学生综合信息员工作细则"; "关于组建北京外国语大学学生舆情信息员队伍的通知."

69 "长沙医学院维稳信息报送制度暂行规定"; "湖南工学院二级学院维稳综治安全信息报送管理办法."

70 Hubei University of Economics, "艺术设计学院信息员队伍建设办法," June 15, 2017, http://ysxy.hbue.edu.cn/index.php/View/329.html; University of South China, "建筑学院学生信息员队伍建设办法," December 28, 2018, https://jzxy.usc.edu.cn/info/1027/1214.htm.

71 For specific provisions on recruitment of student informants, see "关于组建北京外国语大学学生舆情信息员队伍的通知"; "长沙医学院维稳信息报送制度暂行规定"; "湖南工学院维稳综治安全信息报送管理办法"; "华南农业大学学生综合信息员工作细则."

72 "长沙医学院维稳信息报送制度暂行规定."

73 "湖南工学院二级学院维稳中治安全信息报送管理办法."

74 "长沙医学院维稳信息报送制度暂行规定"; "关于组建北京外国语大学学生舆情信息员队伍的通知."

75 "长沙医学院维稳信息报送制度暂行规定"; "湖南工学院二级学院维稳中治安全信息报送管理办法."

76 "华南农业大学学生综合信息员工作细则"; *合肥工业大学年鉴 2004*, 246.

77 "关于组建北京外国语大学学生舆情信息员队伍的通知"; "华南农业大学学生综合信息员工作细则"; "湖南工学院二级学院维稳中治安全信息报送管理办法"; "长沙医学院维稳信息报送制度暂行规定"; "湖北经济学院, 艺术设计学院信息员队伍建设办法."

78 *合肥工业大学年鉴 2002*, 189; *山东大学年鉴1998*, 231.

79 Valuable research concerning state censorship and state distribution of misinformation in China includes Margaret Roberts, *Censored* (Princeton, NJ: Princeton University Press, 2018); and Gary King, Jennifer Pan, and Margaret E. Roberts, "How the Chinese Government Fabricates Social Media Posts for Strategic Distraction, Not Engaged Argument," *American Political Science Review* 111, no. 3 (2017): 484–501.

80 Rogier Creemers, "Cyber China: Upgrading Propaganda, Public Opinion Work and Social Management for the Twenty-First Century," *Journal of Contemporary China* 26, no. 103 (2017): 85–100; Wen-Hsuan Tsai, "How 'Networked Authoritarianism' Was Operationalized in China," *Journal of Contemporary China* 25, no. 101 (2016): 731–744.

81 *陕西年鉴 2015*, 53; *郴州年鉴 2018*, 110; *阿尔山年鉴 2014–2015*, 180; *伊通年鉴 2015*, 87.

82 *陇南年鉴 2020*, 186.

83 *阿尔山年鉴 2014–2015*, 180.

84 The Chinese name of the police bureaucracy responsible for surveillance of public information networks is "公共信息网络安全监察处." *北京公安年鉴 2001*, 115.

85 *山东省志: 公安志* (1986–2005), 141.

86 The Chinese name of the municipal center for "monitoring and controlling public network security" is 公共信息网络安全监控中心." *北京公安年鉴 2002*, 112.

87 *延安年鉴 2012*, 153.

88 "舒城县公安局公共信息网络安全监察大队," August 11, 2021, http://www.shucheng.gov.cn/public/6598681/29163461.html.

89 *郯城年鉴 2011–2014*, 193; *个旧年鉴 2017*, 217; *北京公安年鉴 2003*, 115.

90 *三台年鉴 2018*, 85; *阿尔山年鉴 2014*, 52; *鄂伦春自治旗年鉴 2015*, 106.

91 额尔古纳市公安局机关简介, http://gk.eegn.gov.cn/?thread-7801-1.html.

92 "北京收钱删帖利益链曝光 一名网警受贿百万落网," *新京报*, March 26, 2014, http://www.xinhuanet.com/video/2014-03/26/c_11994

1976.htm; "网警贿赂网警:替领导删帖," *钱江晚报*, April 18, 2014, http://tech.sina.com.cn/i/2014-04-18/14279330265.shtml.

93 *筠连年鉴 2018*, 75.

94 *天津信息化年鉴 2006*, 224.

95 *贵阳白云年鉴 2017*, 271; *云岩年鉴 2017*, 174; *沁水年鉴 2018*, 200; *内江市市中区年鉴 2017*, 131.

96 *泸县年鉴 2018*, 64; *沁水年鉴 2018*, 200; *云岩年鉴 2017*, 174.

97 Creemers, "Cyber China," 95–96.

98 Cyberspace Administration of China, "互联网用户账号名称管理规定," February 4, 2015, http://www.cac.gov.cn/2015-02/04/c_1114246561.htm.

99 国务院办公厅, "国务院办公厅关于进一步加强互联网上网服务营业场所管理的通知," September 30, 2016, http://www.gov.cn/zhengce/content/2016-09/30/content_5114029.htm.

100 *个旧年鉴 2017*, 217; *合肥年鉴 2014*, 148; *延安年鉴 2019*, 167.

101 *山东省志，公安志 1986–2005*, 144.

102 *河东年鉴 2007*, 159; *贵阳年鉴 2008*, 160; *迪庆年鉴 2014*, 232.

103 *贵阳白云年鉴 2015*, 163.

104 王刚，*基层公安机关网络安全保卫理论与务实*（成都：四川大学出版社, 2013), 45.

105 Police in Changsha claim to have implemented public Wi-Fi monitoring in 2009. *长沙年鉴 2010*, 147.

106 *武汉公安年鉴 2015*, 53.

107 *云岩年鉴 2017*, 174; *遂宁年鉴 2018*, 231; *通江年鉴 2019*, 144.

108 *西盟年鉴 2018*, 170.

109 *四川乐山市公安局文件汇编 3*, 22; *陇县年鉴 2017*, 111.

110 *衡阳年鉴 2019*, 155; *鄂伦春自治旗年鉴 2016*, 118; *沧州市运河区年鉴 2017*, 172.

111 *稷山年鉴 2019*, 35, 108; *郯城年鉴 2011–2014*, 7, 193; *个旧年鉴 2013*, 58, 211.

112 *禄劝年鉴 2016*, 197; *东川年鉴 2013*, 204; *贵阳年鉴 2008*, 161.

113 "内江市公安局狠抓网上重点人员管控工作取得显著成效," https://chinadigitaltimes.net/chinese/134942.html.

7. Upgrading Surveillance

1 For a brief survey of China's surveillance technologies, see Samantha Hoffman, "China's Tech-Enhanced Authoritarianism," *Journal of Democracy* 33, no. 2 (2022): 76–89.

2 Dahlia Peterson, "Foreign Technology and the Surveillance State," in *China's Quest for Foreign Technology: Beyond Espionage,* ed. William Hannas and Didi Kirsten Tatlow, 241–257 (London: Routledge, 2020).

3 Richard Berk, "Artificial Intelligence, Predictive Policing, and Risk Assessment for Law Enforcement," *Annual Review of Criminology* 4, no. 1 (2021): 209–237.

4 Minxin Pei, "Grid Management: China's Latest Institutional Tool of Social Control," *China Leadership Monitor,* no. 67 (2021), https://www.prcleader.org/_files/ugd/af1ede_e105c71ab91640f295f7992ceb1ededb.pdf.

5 The National Crime Information Center, used by federal, state, and local law enforcement in the United States, is a digital compilation of criminal justice information such as criminal records, fugitives, stolen property, and missing persons. See National Crime Information Center, Federal Bureau of Investigation, https://fas.org/irp/agency/doj/fbi/is/ncic.htm.

6 "金盾工程战果辉煌," *中国计算机报*, February 17, 2003, B4; 金卡工程杂志社, "2001–2002 年中国金盾工程情况调查," *金卡工程*, no. 2 (February 2003): 34. Henan, for example, launched phase two of its Golden Shield in 2009. *河南信息化年鉴 2009–2010*, 249.

7 刘静，"何谓金盾工程," *人民公安*, 9 (1999): 40–42.

8 Golden Shield incorporates eight national databases: National Population Basic Information (全国人口基本信息资源库), National Border Entry and Exit Information (全国出入境人员资源库), National Motor Vehicle Drivers Information (全国机动车/驾驶人信息资源库), National Police Officer Basic Information (全国警员基本信息资源库), National Fugitives Information (全国在逃人员信息资源库), National Criminals Information (全国违法犯罪人员信息资源库), National Stolen Motor Vehicles Information (全国被盗抢汽车信息资源库), and National Key Safety Units Information (全国安全重点单位信息资源库). "公安部金盾工程一期建设基本完成 利用信息破案占两成," http://www.gov.cn/gzdt/2005-11/30/content_113209.htm.

9 *广东科技年鉴 2006*, 187.

10 *上海信息化年鉴 2003*, 169–170.

11 The Chinese names of these applications are "国保情报信息管理系统, 旅馆业治安管理信息系统, 公共信息网络安全监察信息网络报警处置系统, 涉毒人员管理信息系统, 全国邪教案件管理分析系统, 外国人管理系统 (境外人员管理)." *九江年鉴 2006*, 145; *广东科技年鉴 2003*, 218; *甘肃信息年鉴 2006*, 156; *黔西南年鉴 2006*, 126; *文山年鉴 2004*, 103.

12 *上海信息化年鉴 2007*, 138.

13 *广东科技年鉴 2010*, 262.

14 *江苏信息化年鉴 2005*, 440.

15 李润森，“开拓进取，科技强警,” *公安研究*, no. 4 (2002): 5–12.

16 See also Sonali Chandel et al., “The Golden Shield Project of China: A Decade Later—An In-Depth Study of the Great Firewall,” in *2019 International Conference on Cyber-Enabled Distributed Computing and Knowledge Discovery* (Piscataway, NJ: IEEE, 2019), 111–119.

17 *江苏年鉴 2007*, 212.

18 The official title of the document in Chinese is “关于开展城市报警与监控技术系统建设的意见,” issued by the MPS on August 25, 2005, http://www.e-gov.org.cn/article-82800.html. The document is not publicly available.

19 公安部, “关于印发 ‘关于深入开展城市报警与监控系统应用工作的意见’ 的通知,” http://www.21csp.com.cn/html/View_2011/06/21/4946133222.shtml.

20 戴林, “3111 试点工程建设中监控报警联网系统设计要点分析,” *中国安防产品信息*, no. 4 (2006): 15.

21 戴林，“3111 试点工程建设中监控报警联网系统设计要点分析.”

22 黄海军，“新平安城市建设需要什么?” *中国安防*, no. 3 (2013): 71–74.

23 贺小花，“公安视频监控建设现状,” *第14届安博会* (2013): 123.

24 *黄平年鉴 2017*, 89.

25 *武汉公安年鉴 2013*, 70; *武汉公安年鉴 2014*, 73.

26 *瓮安年鉴 2016*, 173.

27 曲晓顺, “加强基层情报信息工作的探索,” *网络安全技术与应用*, no. 2 (2011): 5.

28 *湖南年鉴 2020*, 475.

29 This addition was given its own name, the Law and Public Order Video Image System (社会治安视频系统).

30 “浏阳市公安局天网工程专项资金绩效评价报告,” http://www.liuyang.gov.cn/lyszf/xxgkml/szfgzbm/sczj/tzgg/201712/t20171220_6481522.html. Changsha launched its fourth phase in 2020 as well. “24小时在岗的警察！长沙 ‘天网工程’（四期）项目建设开工,” *澎湃新闻*, https://www.thepaper.cn/newsDetail_forward_6049308.

31 Ten million people lived in Changsha in 2020, according to that year’s census. “长沙市第七次全国人口普查公报,” June 21, 2021, https://hn.rednet.cn/content/2021/06/21/9571662.html.

32 澎湃新闻, “24小时在岗的警察！长沙 ‘天网工程’（四期）项目建设开工.”

33 广东省公安厅科技处,"广东省治安防控体系建设," http://nj2008.21csp.com.cn/yhp/d2oz/d4j/d4j.htm.

34 高勇，"高屋建瓴," *中国安防*, no. 10 (2014): 2–7.

35 *成华年鉴 2015*, 97; *成华年鉴 2016*, 130; *成华年鉴 2017*, 141; *成华年鉴 2018*, 135; *成华年鉴 2019*, 133; *成华年鉴 2020*, 153.

36 公安部科技局安全技术防范工作指导处, "城市报警与监控系统建设工作进展," http://nj2007.21csp.com.cn/cxp/2/cx-2.2.htm. The PSB of Chenghua District, Chendgu, leased surveillance equipment from Chengdu Telecom. *成华年鉴 2015*, 97; 吴胜益，徐超帅，"江西天网工程在公安办案中应用效果的研究," *科技广场*, no. 12 (2014): 242.

37 许秀燕，"天网工程新型社会治安防控体系," *Informatization of China's Construction,* no. 9 (2019): 35.

38 丁家祥，"城市社会治安图像监控系统的现状与发展趋势分析," *公安研究*, no. 7 (2008): 77.

39 胡海，"公安机关天网工程建管用中存在的问题及解决方案探究," *China's New Technologies and Products,* no. 11 (2015): 26. Local governments in China have few options for raising revenue and rely heavily on proceeds from sales of land-use rights, which fluctuate with overall economic conditions. Yuanyan Sophia Zhang and Steven Barnett, "Fiscal Vulnerabilities and Risks from Local Government Finance in China," IMF Working Paper WP/14/4, International Monetary Fund, January 2014.

40 马云鹏等，"公安视频侦查体系化建设若干思考与建议," *中国刑警学院学报*, no. 1 (2017): 37–44.

41 郎江涛，"公安系统天网工程瓶颈及未来展望," *Science and Technology & Innovation,* no. 9 (2017): 45.

42 马云鹏等，"公安视频侦查体系化建设若干思考与建议," 37–44; 万程，何毅, "长沙市视频监控'天网工程'建设现状及存在问题分析," *考试周刊*, no. 75 (2017): 189–190; 吴胜益，徐超帅, "江西天网工程在公安办案中应用效果的研究": 243–244.

43 中共中央办公厅国务院办公厅, "关于加强社会治安防控体系建设的意见," http://www.gov.cn/gongbao/content/2015/content_2847873.htm.

44 井立国，"雪亮工程示范," *法制与社会* no. 10 (2020): 134; *贵州年鉴 2019*, 94; *奉贤年鉴 2019*, 126.

45 "关于加强公共安全视频监控建设联网应用工作的若干意见," http://news.21csp.com.cn/c23/201505/82143.html.

46 "中共中央 国务院关于实施乡村振兴战略的意见," http://www.gov.cn/zhengce/2018-02/04/content_5263807.htm; 丁兆威, "雪亮工程照亮平安乡村路," *第九届深圳国际智能交通与卫星导航位置服务展览会*

(2020): 126. The first national conference on the buildout of Sharp Eyes was held in October 2016. Another national conference was convened to promote the project in June 2017. 丁兆威，“雪亮工程照亮平安乡村路,” 126.

47 张艳华，“如何建设雪亮工程,” *第七届深圳国际智能交通与卫星导航位置服务展览会* (2018): 137.

48 龙鹏宇，“重庆市永川区搭建平安综治云,” *重庆行政* 22, no. 4 (2021): 61; *永城 年鉴 2020*, 355; *莒县年鉴 2020*, 146.

49 “唐河县公共安全视频监控建设联网应用项目,” https://www.faanw.com/xueliangongcheng/507.html, accessed January 3, 2022; 咸宁市自然资源和规划局, “关于加强全市城镇住宅小区公共安全视频监控设施建设和管理办法,” http://zrzyhghj.xianning.gov.cn/hdjl/dczj/201912/t20191227_1898675.shtml.

50 “西安市公共安全视频图像信息系统管理办法（草案）征求意见稿,” http://www.xa.gov.cn/ptl/def/def/index_1121_6774_ci_trid_2771909.html; “雪亮工程的一、二、三类点怎么区分的?” https://www.jimay.com/support/375.html.

51 “2021 年雪亮工程视频监控系统解决方案(接入公安天网),” https://www.jimay.com/solutions/2963.html.

52 *中卫年鉴 2020,* 154.

53 “西安市公共安全视频图像信息系统管理办法（草案）征求意见稿,” http://www.xa.gov.cn/ptl/def/def/index_1121_6774_ci_trid_2771909.html.

54 *厦门年鉴 2019*, 134.

55 *南丰年鉴 2018*, 153; *濮阳年鉴 2020*, 107.

56 黄松涛，盛进，“雪亮工程检测常见问题,” *中国安防*, no. 8 (2020): 11–13.

57 陈杰，“新时期雪亮工程建设下安防及信息化技术运用,” *网络安全技术与应用*, no. 7 (2021): 144–146; 龙鹏宇, “重庆市永川区搭建平安综治云”: 61–63.

58 杨淼, “视频监控点位规划的实践与思考,” *人民法治* (June 2017): 86.

59 The first mention of a social credit system appeared in the resolution of the Third Plenum of the Eighteenth CCP Central Committee in November 2013. http://www.gov.cn/jrzg/2013-11/15/content_2528179.htm.

60 国务院，“社会信用体系建设规划纲要, (2014–2020年),” http://www.gov.cn/zhengce/content/2014-06/27/content_8913.htm.

61 “国务院办公厅关于加快推进社会信用体系建设构建以信用为基础的新型监管机制的指导意见,” http://www.gov.cn/zhengce/content/2019-07/16/content_5410120.htm.

62 何玲，“以信筑城：第三批社会信用体系建设示范区观察，” *中国信用*, no. 12 (2021): 15–27.

63 “国务院办公厅关于进一步完善失信约束制度 构建诚信建设长效机制的指导意见，” http://www.gov.cn/zhengce/content/2020-12/18/content_5570954.htm.

64 中共中央办公厅 国务院办公厅印发 关于推进社会信用体系建设高质量发展促进形成新发展格局的意见, http://www.gov.cn/zhengce/2022-03/29/content_5682283.htm.

65 Fan Liang, Vishnupriya Das, Nadiya Kostyuk, and Muzammil M. Hussain, “Constructing a Data-driven Society: China's Social Credit System as a State Surveillance Infrastructure,” *Policy & Internet* 10, no. 4 (2018): 415–453; Katja Drinhausen and Vincent Brussee, “China's Social Credit System in 2021: From Fragmentation towards Integration,” *MERICs China Monitor,* Mercator Institute for China Studies, Berlin, March 3, 2021, updated May 9, 2022, 1–24; Xu Xu, Genia Kostka, and Xun Cao, “Information Control and Public Support for Social Credit Systems in China,” *Journal of Politics* 84, no. 4 (2022): 2230–2245.

66 国务院, “社会信用体系建设规划纲要 (2014–2020年).”

67 中国执行信息公开网, “全国法院失信被执行人名单信息公布与查询平台,” http://zxgk.court.gov.cn/shixin.

68 Local regulations on social credit seem to be written using the same template and thus often contain similar provisions. Credit China, the official national website, tracks many local regulations on social credit. *信用中国*, https://www.creditchina.gov.cn/zhengcefagui/?navPage=2.

69 “上海市人大常委会通过全力做好疫情防控工作决定,” *新华网*, February 12, 2020, http://www.npc.gov.cn/npc/c30834/202002/dd8a5ee6fbaf4194bc71a636ab7c5600.shtml.

70 “广东省社会信用条例,” http://www.gd.gov.cn/zwgk/wjk/zcfgk/content/post_2718326.html; 国家互联网信息办公室关于《互联网信息服务严重失信主体信用信息管理办法》(征求意稿), http://www.cac.gov.cn/2019-07/22/c_1124782573.htm.

71 “南京市社会信用条例,” https://www.creditchina.gov.cn/zhengcefagui/xinyonglifa/202101/t20210103_222439.html.

72 “涉疫信息隐瞒不报，巩义这七个人被列入失信 黑名单,” *信用中国*, http://www.zqdh.gov.cn/zwgk/ztzl/xyzggdzqdh/fxts/content/post_2101592.html.

73 “北京：电动自行车进楼道充电，将影响个人征信,” http://kfqgw.beijing.gov.cn/zwgkkfq/zcfg/hygq/202202/t20220217_2611128.html; “该给征信滥用亮起‘红灯’了,” *北京青年报*, December 28, 2020, http://www.xinhuanet.com/comments/2020-12/28/c_1126914404.htm; “滥用个人征信，是

对信用社会失信," *新京报*, April 19, 2019, https://m.bjnews.com.cn/detail/155568606514920.html.

74 "常州市信访人信用管理实施办法(试行)," *中新网*, October 24, 2018, http://www.js.chinanews.com.cn/news/2018/1024/183745.html; "多地出台文件惩罚失信访民 学者称于法无据," *财新网*, September 12, 2019, https://china.caixin.com/2019-09-12/101461655.html.

75 Wen-Hsuan Tsai, Hsin-Hsien Wang, and Ruihua Lin, "Hobbling Big Brother: Top-Level Design and Local Discretion in China's Social Credit System," *China Journal* 86, no. 1 (2021): 1–20.

76 贺凤，"社会信用体系建设中存在的问题及难点," *北方金融*, no. 2 (2021): 108; Chen Huirong and Sheena Greitens, "Information Capacity and Social Order: The Local Politics of Information Integration in China," *Governance* 35, no. 2 (2022): 497–523.

Acknowledgments

I am deeply indebted to many colleagues and friends, whose support made this book possible. Jean Hung and Celia Chan at the University Service Center at the Chinese University of Hong Kong hosted me during my many visits and helped me find materials in the center's collection. At Claremont McKenna College, my able and diligent research assistants Lucy Deng, Genevieve Collins, and Carley Barnhart located sources and summarized literature. Andrew Nathan and Teng Biao helped me set up interviews with exiled dissidents whose encounters with the Chinese surveillance state enriched my understanding of its tactics. Song Yongyi and Chris Buckley generously shared valuable materials and useful leads.

Some of the chapters were presented in seminars at the Keck Center for International and Strategic Studies at Claremont McKenna. I am grateful for the comments and suggestions of my colleagues. I also presented some of the initial findings at the Hoover Institution at Stanford University, where I received helpful comments from Larry Diamond, Jean Oi, and Glenn Tiffert.

I am especially indebted to Andrew Walder, who read the entire manuscript and offered invaluable guidance. I also want to thank Nancy Hearst, my longtime copyeditor, for cleaning up my writing and catching an embarrassingly large number of errors.

I am grateful to the Smith Richardson Foundation for funding my research. In particular, I thank Marin Strmecki and Allan Song for steadfastly supporting me over the past two decades.

I deeply appreciate the helpful comments and criticisms from the three anonymous reviewers. Finally, I want to thank Kathleen McDermott, my editor at Harvard University Press, for her encouragement and confidence in the project book. I also deeply appreciate Simon Waxman's meticulous editing, which has made this book cleaner and clearer.

This book is dedicated to the memory of my elder sister Pei Xingmei, whose care, protection, and love helped me survive the Cultural Revolution.

Index